MLOps with Ray

Best Practices and Strategies for Adopting Machine Learning Operations

Hien Luu
Max Pumperla
Zhe Zhang

Apress®

MLOps with Ray: Best Practices and Strategies for Adopting Machine Learning Operations

Hien Luu
Santa Clara, CA, USA

Max Pumperla
Bad Segeberg, Germany

Zhe Zhang
Sunnyvale, CA, USA

ISBN-13 (pbk): 979-8-8688-0375-8
https://doi.org/10.1007/979-8-8688-0376-5

ISBN-13 (electronic): 979-8-8688-0376-5

Managing Director, Apress Media LLC: Welmoed Spahr
Acquisitions Editor: Celestin Suresh John
Development Editor: Laura Berendson
Coordinating Editor: Gryffin Winkler

Cover designed by eStudioCalamar
Cover image designed by Freepik (www.freepik.com)

Distributed to the book trade worldwide by Apress Media, LLC, 1 New York Plaza, New York, NY 10004, U.S.A. Phone 1-800-SPRINGER, fax (201) 348-4505, e-mail orders-ny@springer-sbm.com, or visit www.springeronline.com. Apress Media, LLC is a California LLC and the sole member (owner) is Springer Science + Business Media Finance Inc (SSBM Finance Inc). SSBM Finance Inc is a **Delaware** corporation.

For information on translations, please e-mail booktranslations@springernature.com; for reprint, paperback, or audio rights, please e-mail bookpermissions@springernature.com.

Apress titles may be purchased in bulk for academic, corporate, or promotional use. eBook versions and licenses are also available for most titles. For more information, reference our Print and eBook Bulk Sales web page at http://www.apress.com/bulk-sales.

Any source code or other supplementary material referenced by the author in this book is available to readers on GitHub (https://github.com/Apress). For more detailed information, please visit https://www.apress.com/gp/services/source-code.

If disposing of this product, please recycle the paper

Table of Contents

About the Authors

Hien Luu is a passionate AI/ML engineering leader who has been leading the Machine Learning platform at DoorDash since 2020. Hien focuses on developing robust and scalable AI/ML infrastructure for real-world applications. He is the author of the book *Beginning Apache Spark 3* and a speaker at conferences such as MLOps World, QCon (SF, NY, London), GHC 2022, Data+AI Summit, and more.

Max Pumperla is a data science professor and software engineer located in Hamburg, Germany. He's an active open source contributor, maintainer of several Python packages, and author of machine learning books. He currently works as a software engineer at Anyscale. As the head of product research at Pathmind Inc., he was developing reinforcement learning solutions for industrial applications at scale using Ray RLlib, Serve, and Tune. Max has been a core developer of DL4J at Skymind and helped grow and extend the Keras ecosystem.

Zhe Zhang has been leading the Ray Engineering team at Anyscale since 2020. Before that, he was at LinkedIn managing the Big Data/AI Compute team (providing Hadoop/Spark/TensorFlow as services). He has been working on open source for about a decade. Zhe is a committer and PMC member of Apache Hadoop; he is the lead author of the HDFS Erasure Coding feature, which is a critical part of Apache Hadoop 3.0. In 2020, Zhe was elected as a Member of the Apache Software Foundation.

About the Technical Reviewer

 **Ashutosh Parida** is an accomplished leader in Artificial Intelligence and Machine Learning (AI/ML) and currently serving as Assistant Vice President, heading AI/ML product development at DeHaat, a leading AgriTech startup in India. With over a decade of experience in data science, his expertise spans various domains, including vision, NLU, recommendation engines, and forecasting.

With a bachelor's degree in Computer Science and Engineering from IIIT Hyderabad and a career spanning 18 years at global technology leaders like Oracle, Samsung, Akamai, and Qualcomm, Ashutosh has been site lead for multiple projects and has launched products serving millions of users. He also has open source contributions to his credit.

Stay connected with Ashutosh on LinkedIn to stay updated on his pioneering work and gain valuable industry insights: linkedin.com/in/ashutoshparida.

CHAPTER 1

Introduction to MLOps

Machine learning (ML) has proven to be a very powerful tool to learn and extract patterns from data. The ability to generate, store, and process a large amount of data, and easily access computing power in the last decade has contributed to many advancements in the ML field, such as image recognition, language translation, and large language models (LLMs), that is, BERT, DALLE, ChatGPT, and more.

ML has finally graduated from the academia lab and has been embraced with both open arms by the business world to help with solving real-world business problems and transforming industries by improving customer experience, reducing cost, improving business efficiency, and ultimately increasing their competitive advantage. According to McKinsey's "The state of AI in 2021"[1] report, the findings from the survey indicate that AI/ML adoption is continuing its steady rise across many companies in many regions of the world. One of the reasons for this rise is due to the impact that AI has on the business bottom line.

It is no longer a question whether AI/ML can deliver business values to organizations across the industries. A more pressing question on an AI/ML leadership team's mind nowadays is how can their organizations most efficiently integrate AI/ML into their business processes or products to deliver that value. To do so, they need the ability to operationalize ML in an iterative, consistent, effective, safe, and predictable manner.

The data about the number of ML project failures from various sources such as Gartner[2] and VentureBeat[3] suggests that operationalizing ML is a complex endeavor that requires a set of standardized processes and technology capabilities for building, deploying, and operationalizing ML models efficiently and quickly. Welcome to MLOps!

---

[1] Global Survey: The State of AI Adoption 2021 – www.mckinsey.com/capabilities/quantumblack/our-insights/global-survey-the-state-of-ai-in-2021

[2] Our Top Data and Analytics Predicts for 2019 – https://blogs.gartner.com/andrew_white/2019/01/03/our-top-data-and-analytics-predicts-for-2019/

[3] Why do 87% of data science projects never make it into production? – https://venturebeat.com/ai/why-do-87-of-data-science-projects-never-make-it-into-production/

© Hien Luu, Max Pumperla and Zhe Zhang 2024
H. Luu et al., *MLOps with Ray*, https://doi.org/10.1007/979-8-8688-0376-5_1

MLOps Overview

Software engineering projects are designed to bring values to a company, organization, or a team. The realization of the intended values will only start when the development software artifacts are successfully deployed into production. The sooner this happens the better. Companies around the world have adopted DevOps as methodologies to reliably develop and deploy large-scale software to production by minimizing the gap between development and operations and promoting collaboration, communication, and knowledge sharing. DevOps is widely adopted in the industry because it helps with fast, frequent, and reliable software releases by emphasizing automation with continuous integration, continuous delivery, and continuous deployment.

Similarly, ML projects are designed to add certain value to a company, organization, or a team. The ROI of the ML projects will only start when the ML artifacts, the ML model and features, are deployed to production and properly monitored. The achievement of the ROI for ML projects can vary due to factors such as the project's complexity, data quality, and specific goals. Organizations may start seeing initial returns during the early stages of deployment, especially if the ML models are addressing specific pain points like reducing churn in customer subscriptions. The full ROI is typically realized when models are successfully integrated into operational processes, delivering sustained and measurable value over an extended period. Achieving this often takes time, as models may need refinement, optimization, and continuous monitoring to adapt to changing conditions.

How are ML projects different from the software engineering projects? What's unique about ML projects? Can DevOps methodologies help with ML projects? Let's examine these questions to deeply understand MLOps and the benefits it provides.

Note   DevOps has been widely adopted across many organizations that are developing software as methodologies to improve software quality and reliability while reducing time to market for software engineering initiatives. It is both a paradigm shift to address social and technical issues in organizations engaged in software development and a continuous process of automation across the software development process.

The cornerstone of DevOps is a continuous process including continuous development, integration, deployment, and monitoring that is designed for fast, frequent, and reliable software releases.

The DevOps mindset requires software engineers to care not only about what they develop, but also care about the software deployment and operations.

ML Projects

Similar to software engineering projects, ML projects have their own development lifecycle. However, their lifecycle is a highly iterative one due to the scientific nature of the ML model training process, which requires experimentation and has a large dependency on the quality of the data used to train the ML algorithm. As a result, the ML development process is not a linear one, like the standard software engineering development lifecycle, but rather it is cyclical in that it demands iteration, tuning, and improvement, as depicted in Figure 1-1.

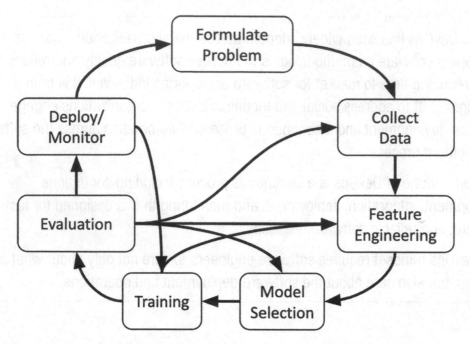

Figure 1-1. *ML development process*

ML projects often are initiated to help fulfill certain business or product initiatives with measurable outcomes. As the first step of the ML project, it is important to formulate the problem with clear goals and objectives before going through the other steps in the clockwise manner. It is not unusual, but rather expected and necessary that data scientists will need to go back to previous steps, such as to collect additional data or change the way features are generated, if the model evaluation step indicates the ML model doesn't perform at the expected level or when the experiment results convince data scientists to change the current approach or continue to fine-tune the current approach.

Successful ML projects are the ones where data scientists are able to make progress throughout the ML development lifecycle as quickly as possible and able to go through the development loop as many times as needed in order to apply the insights from previous experiments to fine-tune their approach, the needed data, and ultimately produce the most optimal ML models. The ultimate goal is to produce the most optimal ML models that can perform predictions with new data at the accuracy level to meet business goals or objectives.

Although the ML development lifecycle is cyclical and appears complex, it can be simplified into five phases:

- Data collection and preparation
- Feature engineering
- Model training
- Model deployment
- Model monitoring

ML Project Inputs and Artifacts

In traditional software engineering projects, software engineers write code to develop logic or algorithms to meet the given specifications to produce the output based on the input, as depicted in Figure 1-2.

Figure 1-2. *Software projects*

In ML projects, data scientists spend a lot of time and focus on these two main activities: feature engineering and model development. The following paragraphs are meant to briefly describe these two activities and more importantly to call out the ML project inputs and artifacts.

The quantity and quality of the ML features have a huge contribution to the performance of the ML models. Data scientists typically spend a large portion of their time in collecting and analyzing the collected data, and then write code to transform the data into ML features to train the ML models with, as depicted in Figure 1-3.

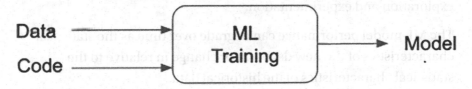

Figure 1-3. *Model training in ML projects*

Once data scientists are reasonably happy with the generated ML features, they then proceed to the ML model training step, which involves writing code to train ML models using the generated features, an ML algorithm, and a set of tuning parameters. This step often requires exploration and experimentation to evaluate, fine-tune, and iterate on the ML model to improve its performance. If the model evaluation results are suboptimal, data scientists might need to go back to the feature engineering step to collect more or different sets of data, or select a different ML algorithm, etc.

The main ML artifacts from the above activities are as follows:

- The data used to generate features

- The logic for generating features from the collected data

- The code and parameters for training the ML algorithm

- The produced ML model

Oftentimes, the ML models will need to be retrained because new business requirements emerge, or additional new data sources or ML libraries are available, or the performance of ML models has started to degrade, and more. Hence, it is important to version control or manage the ML artifacts listed previously and depicted in Figure 1-4.

ML Model = Data + ML Algorithm + Hyperparameters

Figure 1-4. *ML artifacts*

ML projects are different traditional software engineering projects and present many unique challenges and can be summarized in the following points:

- Training ML models requires historical data, and hence ML projects involve more data-related activities, such as collecting and labeling data and analyzing and visualizing input data to better understand its statistical characteristics.

- Model development is a high iterative process that requires exploration and experimentation.

- The ML model performance can degrade over time as the statistical characteristics of the new data drift or change in relative to the statistical characteristics of the historical data.

- ML projects typically require more collaboration between data scientists, data engineers, ML engineers, and domain experts, as the success of the project often depends on a combination of technical expertise and domain knowledge.

It has been said that MLOps is to machine learning, as DevOps is to software engineering. The main goal of MLOps is to help companies around the world to accelerate their ML projects to production in a repeatable, consistent, and efficient manner by prescribing a set of best practices around both the technical and non-technical elements.

MLOps: The Missing Element

As more and more companies around the world recognize the power of AI/ML and allocate budget to invest in applying AI/ML to add business values, to increase their competitiveness, and more, they rightly so want to see the ROI of their investment in ML projects. Their ROI can only start when the ML models are deployed to production and integrate into their products or business processes. What can ultimately help with bringing ML models to production efficiently and quickly? From the collective wisdom and learning, the ML practitioner community and industry have come up with the answer, which is called MLOps.

In this section, we will examine some of the common challenges and pitfalls in operationalizing ML projects, and then discuss at the high level how MLOps can help with those with the goal of improving ML project success.

Operationalize ML Project Challenges

As reported by numerous surveys from Gartners and NewVantage Partners, many organizations are not as successful as they would like to be when it comes to getting the ROI from their investment in ML projects due to the challenges of operationalizing ML models in production quickly, efficiently, and consistently. According to a survey conducted by NewVantage Partners in 2020,[4] only about 15% of leading organizations have deployed AI capabilities into production at any scale.

---

[4] AI Stats News: Only 14.6% Of Firms Have Deployed AI Capabilities In Production –www.forbes.com/sites/gilpress/2020/01/13/ai-stats-news-only-146-of-firms-have-deployed-ai-capabilities-in-production/

It is fair to say most organizations now understand operationalizing ML projects is challenging and it is not the same as operationalizing traditional software projects.

The section will highlight a few common challenges in ML projects and shed some lights about how those challenges come about. These challenges we believe are a subset of the ones that are contributing to ML project failure. Challenges around a lack of talent or unclear business objectives are beyond the scope of this book.

Applying Machine Learning

In the real-world ML project, it is generally well understood that the central piece of these projects is applying ML to enable data-driven decisions and products.

ML, as a discipline, is highly experimental and iterative in nature, especially compared to typical software engineering.

Applying ML involves training a model on a dataset and then evaluating its performance on the same dataset as well as a separate dataset. Rarely the first version of the model will meet the expected performance, and therefore, this process can be repeated multiple times, where each iteration potentially uses a different ML algorithm or architecture, hyperparameter settings, and feature engineering techniques.

At the beginning of applying ML, it is very challenging to know the exact combination of the aforementioned parts that will lead to an ML model that performs well. Therefore, exploration, experimentation, and iteration are necessary parts of finding the best combination and quickly pruning out the bad ones.

Similar to other scientific endeavors, ML experimentation and iteration require the ability to record the experiment input, the approach, results, and more. This will facilitate and speed up the process of analyzing the result by comparing multiple experimentation runs to determine the next step.

Additionally, ML is still an evolving field, with new techniques, approaches, and ML libraries being developed all the time. As a result, ML practitioners must be open to experiment with and take advantage of these new changes to improve the ML model performance.

The name of the game here is velocity. If ML practitioners are bogged down not having the processes or tools to explore, experiment, and iterate easily and quickly, then getting ML models to production might not be feasible or will take a long time to realize the ROI of the ML project investment.

Garbage In, Garbage Out

The classic saying of "Garbage In, Garbage Out" in computing is about how problematic input data will produce problematic outputs, which is especially relevant to ML, given that the model training process relies heavily on the quality of the input data.

ML model training is the process where the labeled training data is fed into the ML algorithm to learn its pattern. It is widely known the ML model performance is only as good as the quality of that data. Some of the well-known ML practitioners recently started advocating for the data-centric AI approach to further highlight the benefits of high-quality training data. For more details about this approach, see the notes below.

In addition to data quality, other data-related aspects that have large influence on the ML model performance are data freshness and changes to data statistical properties.

If there is a lack of data infrastructure, data engineering rigor, and personnel support around these data-related aspects, then that will have a negative impact on the ML model performance, which ultimately slows down the path to bring ML models to production.

Note   Model-centric AI vs. data-centric AI – same goal, but different approach

Model-centric AI is an approach and mindset in improving ML model performance by focusing on tweaking the hyperparameters or changing the model architecture or algorithms until the desired metrics are achieved. This approach is what traditionally the industry has been practicing.

Data-centric AI has the same goal as the model-centric AI, but it takes a different approach by holding the hyperparameters and the model architecture and algorithm fixed while applying error analysis driven data iteration to improve the model performance. Formally, data-centric AI is defined as the discipline of systematically engineering the data used to build an AI system, according to the Data-centric AI Resource Hub website[5]. This approach was introduced and advocated by Andrew Ng in his "A Chat w/ Andrew on MLOps: From Model-centric to Data-centric AI" video[6].

---

[5] Data-centric AI Resource Hub – https://datacentricai.org/

[6] A Chat with Andrew on MLOps: From Model-centric to Data-centric AI –www.youtube.com/watch?v=06-AZXmwHju

In the Beginning

Traditionally, ML has been approached from a perspective of individual scientific experiments which are predominantly carried out in isolation by data scientists. Data scientists are knowledgeable and trained in the ML field, and are mainly tasked with model training-related tasks.

As a result, data scientists tend to focus less on areas that are peripheral to ML model training area, such as spending time on automating the data pipelines, focusing on developing robust and high-quality code, automating end-to-end model training pipelines, and integrating ML models into the data-driven products in production.

Enterprise ML projects are not about experimenting and developing ML models one time and moving on to other projects; they demand those software engineering-related activities to be automated, version controlled, monitored, reproducible, and more.

Without having data scientists shifting toward a more software engineering-centric mindset or having the necessary tooling, infrastructure, or processes to support data scientists with those software engineering-related activities, then there will be a negative impact on the road to operationalizing ML in your organization. In addition, there needs to be a culture shift toward a product-oriented mindset at the organization level.

Team Sport

The end-to-end sequence of developing ML models and incorporating them into the data-driven decision products is a complex and interdisciplinary process, such as data engineering, machine learning, software engineering, and developer operations. Undoubtedly, it is a team sport, and thus responsibilities and ownership need to be clear and collaboration across teams is required to ensure those products are stable, are kept up to date, and more importantly continue to add value to the business.

A typical team sport consists of role and responsibilities, and given productionalizing ML is branded as a team sport, let's capture some of the typical roles and responsibilities across the activities in the ML development lifecycle.

Table 1-1 is meant to capture the core set of ML development activities and not meant to be comprehensive.

Table 1-1. *ML development
activity assignment*

Activity	Role
Data preparation	Data engineer
Feature engineering	Data scientist
Model training	Data scientist
Model deployment	ML engineer
Model monitoring	ML engineer

Some enterprises might include other roles into the ML operationalization process, such as business stakeholder or data governance officer.

For medium to large enterprises, each of those roles might be carried out by one or more people. For smaller enterprises or startups, two or more or all of those roles are carried by an individual.

A more elaborated depiction of the intersections of the various roles from the "Machine Learning Operation (MLOps): Overview, Definition and Architecture"[7] paper is depicted in Figure 1-5.

---

[7] Machine Learning Operations (MLOps): Overview, Definition, and Architecture – https://arxiv.org/ftp/arxiv/papers/2205/2205.02302.pdf

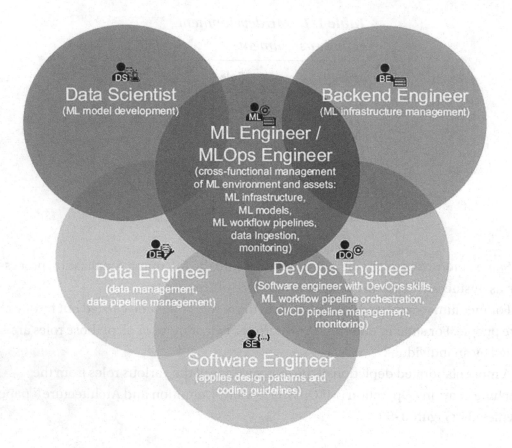

Figure 1-5. *An elaborated depiction of the intersections of various roles in MLOps*

In order to be a winning team in a team sport, it needs to have all the needed personnel and a clear map of responsibilities, and a clear protocol around communication and coordination. Similarly, for enterprises to be successful at operationalizing ML, they need the necessary and sufficient talent that formed a multi-disciplinary team to perform those roles, and there need to be standard processes, clear communication protocols, and boundaries of ownership and responsibilities so all the teams involved are on the same page and aligned, fulfilling the expected deliverables and performing proper handoff at the right time.

Summary of Challenges

It is reasonable to say that successfully operationalizing ML consistently, efficiently, and at scale is not an easy task; however, it is feasible and results are worthwhile.

ML has proven to be a disruptive technology that can help with reducing cost, improving operational efficiency, and increasing business bottom-line for those enterprises that are willing to invest in it.

This section summarizes the challenges listed above in the context of the following three key dimensions, which MLOps aims to address and will be described in the "The Promise" section: automation, reproducibility, and monitoring.

Automation

As described above in the "Applying Machine Learning" section, it is a highly experimental and iterative process. This means that most if not all activities in the process of applying ML can greatly benefit from the automation. Manual process presents many challenges including error prone, time consuming, inconsistency, and not easily reproducible.

Automation leads to increased velocity because those activities can be repeated easily and consistently and enable data scientists to be able to go through the development lifecycle quicker.

As described in the "Garbage In, Garbage Out" section, the data-related activities are quite important to the ML model performance. Automation of the data-related activities such as running and monitoring data pipelines to ensure data quality and freshness will ultimately contribute to the optimal ML model performance.

Reproducibility

Complex ML projects often require collaboration between multiple data scientists to discuss their hypothesis and validating ML model training experiments. To be effective at this, they need the ability to reproduce the experiments easily, and this requires the artifacts that were used in the previous experiments to be readily available and accessible.

It is very common for data scientists to iterate on an existing ML model to create a new one. Common reasons for the new ML model iterations include changing business requirements, new training data sources are available, customer behavior changes, and more. The ability to reproduce most of what was already done in the previous ML model version will greatly speed up the new iteration.

Monitoring

It is often said that deploying ML models into production is only half the battle, and that continuously maintaining their performance afterward is the other half. This is because ML model performance in production can and often does degrade, which can have negative impacts on customers or the enterprise.

This degradation can occur for a variety of reasons, such as quality issues with the ML features used for predictions, changes in user behaviors, changes in the environment (such as the COVID pandemic), and more. Therefore, it is essential to continuously monitor the performance of deployed ML models and alert data scientists when the performance falls below a certain threshold.

Similar rigor and continuous monitoring should also be applied to the data pipelines that produce the data for training or generating features used during prediction time.

By actively monitoring the data, features, and performance of ML models, data scientists and ML engineers can identify issues early and take appropriate action to maintain the effectiveness of the models over time.

MLOps: The Promise

Now that we gain a high-level understanding of the development lifecycle and inputs and artifacts of ML projects, as well as the common challenges in operationalizing ML models, let's examine what MLOps is and how it can help with addressing those challenges in the context of the three dimensions listed in the "Summary of Challenges" section.

A quick search on the Internet will review many slightly different interpretations of MLOps, as well as the definition. However, they all tend to converge toward a common goal and a set of themes.

To understand the essence of MLOps, we will peel back the layers of what it encompasses, similar to peeling an onion. MLOps consists of three fundamental layers: paradigm, engineering discipline, and principle. Each layer contributes to the ultimate goal of MLOps and will be explored in detail. By examining each layer individually and collectively, we will gain a comprehensive understanding of the principles and practices that underpin MLOps, as well as its significance in enabling efficient and effective management of the entire machine learning lifecycle. The MLOps onion is depicted in Figure 1-6.

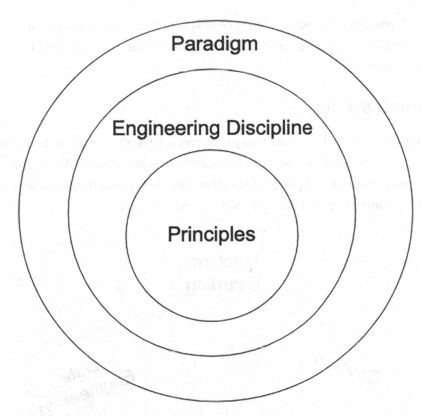

Figure 1-6. *MLOps onion – paradigm, engineering discipline, principles*

Paradigm

MLOps represents a paradigm shift in the way organizations approach machine learning. This paradigm recognizes that machine learning is not just a research or experimental activity but a critical component of business operations that must be managed with the same level of rigor and discipline as other technology systems.

Organizations that are at the forefront of reaping the benefits of their successful ML projects are those that have embraced this paradigm and mindset, treating ML models and artifacts as first-class software components. These organizations have also adopted an operational ML mindset, designing, building, and managing ML systems with a focus on reliability, scalability, and efficiency in line with MLOps principles.

The MLOps paradigm consists of a set of best practices, concepts, and a development culture. These aspects will be described in the "Engineering Discipline" and "Principle" sections.

Engineering Discipline

From the rise of applying ML to business problems across many organizations around the world, MLOps emerged as a new engineering discipline that combines the engineering best practices and principles of three existing disciplines, namely, machine learning, data engineering, and devops, as depicted in Figure 1-7.

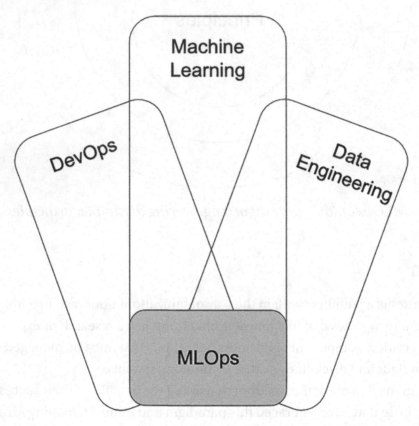

Figure 1-7. *MLOps engineering discipline – intersection of three other disciplines*

MLOps engineering discipline involves the application of engineering principles and practices to the development, deployment, monitoring, and maintenance of machine learning models. Its main goal is to enable organizations to operationalize ML models in an efficient, high velocity, scalable, and maintainable manner. In other words, to reduce friction to get ML models from an idea into production and integrate with software systems in the shortest amount of time with as little risk as possible.

Data Engineering

It is well understood that data is the lifeblood of AI/ML, and the quantity and quality of data will determine the level of performance that ML models can achieve.

Data engineering discipline brings the following main contributions to MLOps discipline:

- Lay the foundation to prepare data that will be used to train ML models.

- Take care of the framework and infrastructure for processing, storage, and consuming data from different data sources in various formats.

- Ensure the data that will be used to train ML models meets the quality standards and freshness requirements via automated, efficient, tested, monitored data pipelines.

Machine Learning

At the heart of applying ML to enterprise business problems is the practice and application of ML techniques.

Machine learning discipline brings the following main contributions to MLOps discipline:

- Analyze and draw insights from the prepared data to determine the data statistical properties to ensure the right level of representation and fairness of the ML model training data.

- Determine and select the most appropriate combination of ML algorithm and tuning parameters to product ML models that will perform well on new data.

- Develop intuition about the ML model performance to iterate and optimize the ML model performance to meet business success metrics before deployment ML models to production.

DevOps

MLOps leverages much of the best practices from well-known and trusted DevOps discipline. There are more artifacts in MLOps than in DevOps, and these additional ones add more complexity and challenges, and they must be treated accordingly in the context of DevOps.

DevOps discipline brings the following main contributions to MLOps discipline:

- Promote collaboration, communication, and knowledge sharing to close the gap between development and operations. Given MLOps is a bigger team sport than DevOps, this contribution is extremely applicable and critical to the MLOps success.

- Ensure automation with continuous integration, delivery, and continuous deployment to drive fast, frequent, and reliable ML model releases. This will help with accelerating the time it takes to go from ideas to ML models in production and maintain their expected performance with ML model retraining when necessary.

- Ensure continuous testing, quality assurance, continuous monitoring, logging, and feedback loops. Given ML model performance depends largely on the training data and new data used for prediction and data changes constantly, therefore continuous monitoring of the data statistical properties and model performance are key to ensuring ML models perform as expected and minimize the risks affecting user experiences or other negative impacts.

The best practices of the three engineering disciplines, data engineering, ML, and DevOps, are a great start; however, some adjustments and additions listed above will need to be incorporated into MLOps due to its unique and highly iterative and experimental nature, being a larger team sport, and artifacts include data, code, and model, which bring additional challenges and management. Below are a few of the notable ones:

- Testing in ML projects is more involved than testing traditional software systems. In addition to unit and integration tests, they also need data validation, trained model quality evaluation, and model validation.

- Deploying ML models involve a multi-step pipeline that includes automatically retrain and deploy model, and deploying features to online feature store on a regular basis.

- Monitoring ML models in production is a must due to the natural tendency that model performance will degrade. This requires regular and proactive tracking of both the summary statistics of the data used for predictions and the online model performance.

- Continuous ML model training is unique to ML systems. This additional practice is the ability to automatically retrain and serve the models with guardrails when there are any changes to the data, code, and ML models.

Principles

In the "Engineering Discipline" section above, there were numerous discussions about some of the best practices in the three disciplines that MLOps is built upon. This section aims to codify some of those and include a few additional ones into a set of principles that any MLOps adoption should consider.

These principles are meant to guide the MLOps practice in an organization. As for the level of rigor of or focus on each principle, that will vary depending on that organization's AI strategy, objective, use cases, and culture.

Automation

Automation refers to the process of removing manual processes and humans involved as much as possible and investing in process and tooling to carry out the critical steps in ML development lifecycle. This includes execute, build, test, train, deploy ML artifacts, such data, code, and ML models.

Some of the ML development activities need to run repeatedly on a certain cadence, such as data pipelines, feature generation pipelines, model training pipelines, and more. These activities will benefit the most from automation.

The need for automation increases

- As the number of ML models reaches a point where manual management becomes impractical and resource-intensive

- As the team member (data scientist, data engineer, ML engineer) size reaches a point where manual coordination and communication become more challenging, such as 10 or more

- As the organization relies more and more on the value that ML models bring to the business

Automation provides fast feedback to participants in the ML project development and thus increases the overall productivity and collaboration.

Versioning

The main artifacts in ML projects are data, code, and ML model. An MLOps best practice is to treat these artifacts as first-class citizens, such as code in the DevOps discipline, using version control systems.

Similar to the best practices for developing software systems, ML model training code should go through a code review process, in addition to versioning, to make the training of ML models auditable and reproducible.

One of the oldest stories in the ML practitioner community is about the inability to reproduce or retrain an ML model because the original ML model author left the company and the training code and metadata were never checked into a version control system. This best practice should be able to help with this.

One of the challenges with versioning the training data is due to its size.

Experiment Tracking

As mentioned before, ML development is a scientific endeavor that is highly iterative and experimental. Supporting this unique aspect of machine learning to enable data scientists to quickly experiment, evaluate, and compare results, collaborating with other data scientists, will require an easy way to track the metadata about the experiment, such as used parameters, performance result metrics, model lineage, data, and code.

The benefits not only include reproducibility, but more importantly the traceability. In addition, ML model exploration and iteration costs both money and time. Therefore, anything that can be done to reduce money and time will bring efficiency to organizations.

Reproducibility

This principle refers to the ability to reproduce the results given the same input for the major steps in the ML development lifecycle, including the feature generation, ML model training and experiments, and ML model deployment.

For traditional software development, using a version control is generally sufficient to satisfy the reproducibility principle. For ML projects, it requires more effort to track the various pipelines, feature generation and ML model training code, hyperparameters, especially data, and the ML model training environment.

The benefits this principle brings are many and vital, such as when an ML project has a change of hands when a data scientist leaves a company or moves on to another project, or when there is a need to debug an important production issue due to negative business impact or regulatory requirements.

Testing

It is fair to expect the practicing of the testing principle should receive the least resistance among data engineers, data scientists, ML engineers, and others. However, due to the dynamic nature and multiple types of inputs and artifacts of ML projects, testing is a bit more challenging as well as even more critical.

Effective testing in ML projects requires multiple types of testing. The following sections focus on two important ML artifacts: data and ML model.

Data-Related Testing

Both the data quality and statistical properties have direct influence on the ML model performance during training and prediction phases. A few important types of testing are listed below.

- Test for null values, abnormal statistical distribution, and feature correlation.

- Validate the assumption about the target label distribution in the case of binary or multi-class classification ML task.

- Unit test the feature generation code.

Model-Related Testing

- Validate the model performance with the offline data to ensure it meets the expected performance metrics such as accuracy, as well as operational metrics like inference latency or model size.

- Generate feature importance to gain insights into the impact of each feature to the model performance

- Perform model smoke test by comparing the model performance of new model or version against a simple baseline model or the current one in production using a small amount of live production data.

- Conduct testing on the model's performance with respect to fairness, bias, and inclusion.

Continuous ML Training, Evaluation, and Deployment

As organizations are looking to scale up their AI/ML investment and ML projects, it is highly recommended to invest in this principle to continuously reap the ROI of their ML projects by keeping ML models continuously adjusted to the dynamic world that we live in.

It is a well-known fact in the ML practitioner community that model performance degrades over time due to many reasons. For ML use cases, such as movie recommendation where the input data changes pretty rapidly due to user behavior changes, it is critical to keep the ML model up to date with continuous ML training and evaluation in order for those ML models to maintain and improve the performance.

Practicing this discipline in an efficient, effective, and safe manner requires the following support:

- Ability to monitor, detect, and alert when model performance starts to fall below a certain acceptable threshold based on business success metrics

- An automatic ML training and evaluation pipeline that is triggered based on data changes or performance degradation

- An automated deployment process to promote and deploy an updated model version in a safe manner

Not all ML use cases require continuous ML training and evaluation. However, the ones that do will greatly benefit from this principle and thus will pay a big dividend to the ML project investment over a long period of time.

Continuous Monitoring

Change is the only constant in life.

—Heraclitus, Greek philosopher

Another unique aspect of ML where ML models require continuous monitoring of their performance to protect the downside when the performance falls below a certain acceptable threshold or starts to produce erroneous predictions that have negative impact to user experience, business value, or organization image.

This principle advocates for regular and proactive assessment across all the ML artifacts: data, model, code, data pipelines, ML model training pipelines, and more, to detect potential errors.

Among the various areas to continuously monitor, the one that is ML related and most challenging is about model drift because this creeps into ML models as time goes by and they can have detrimental effects to business results. This model drift phenomenon refers to a drift in the model's predictions that happens over time as a result of the change in the relationship between what the ML model is trained to predict and the input training data. The section below will share more details and examples to make this clear.

The two types of model drift are concept drift and data drift.

- Concept drift

  - In the ML parlance it means when the functional relationship between the model inputs and outputs changes from a previous time period.

  - An early indicator of this drift is when the model performance is decaying while the statistical distribution of the input features used for prediction stays the same.

- An interesting example of this drift is a grocery item forecast prediction model that was trained before at the beginning of the COVID-19 pandemic. Due to the sudden spike in demand for items like toilet paper and hand sanitizer during the pandemic, the model's predictions for the demand would become increasingly inaccurate.

- Data drift

  - In the ML parlance it means when the statistical properties of the model inputs (input training data or features) changes from a previous time period.

  - An early indicator of this drift is when the model performance is decaying in relation with the changes to the statistical distribution of the input features used for prediction.

  - An interesting example of this drift is the travel time prediction model that was trained before COVID pandemic. The model was no longer effective in predicting travel time during the pandemic because there are less car on the street during the busy hours.

In addition to monitoring model effectiveness, there are other operational metrics related to model serving that also need monitoring, such as

- Model serving latency, a key metric for online ML use cases to indicate the health of model serving system.

- Resource utilization, such as the usage amount of memory, CPUs, GPUs. Excessive usage of these resources suggests there is system instability.

- Model prediction throughput.

- Error rates.

Note   DataOps vs. ModelOps vs. AIOps

There are a few terms that end with Ops and their origin is a bit unclear. The brief definition listed below came from the Gartner glossary section[8] of their website.

DataOps is a collaborative data management practice focused on improving the communication, integration, and automation of data flows between data producers and consumers across an organization.

ModelOps is focused primarily on the governance and lifecycle management of a wide range of operationalized ML models.

AIOps combines big data and ML to automate IT operations processes, including event correlation, anomaly detection, and causality determination.

MLOps Canonical Stack

In addition to MLOps engineering practices and principles, we also need the underpinning technical stack and components to provide a scalable and efficient means to automate and scale the development, deployment, management, and monitoring of ML models, as well as the integration of ML models with product software systems.

This section will discuss the MLOps canonical stack, and it is meant to serve as a blueprint for organizations looking to build out their MLOps infrastructure to operationalize their ML projects. The canonical stack, its components, and core capabilities will be described from an engineering's perspective, not ML, and in a generic and technology-agnostic way.

It is up to organizations to decide on the best-fitting open source and commercially viable technologies and frameworks for their needs, as well as the adoption strategies to maximize the likelihood of success. Much of the adoption strategies and the technical aspects will be discussed in subsequent chapters.

---

[8] Gartner Glossary – www.gartner.com/en/glossary

MLOps Blueprint

The blueprint in Figure 1-8 is an adaptation of the one from the "AI Infrastructure Ecosystem of 2022"[9] report published by the AI Infrastructure Alliance website. Their blueprint is a distillation of the core MLOps capabilities, and this one combines it with a few additional pieces that recently have emerged. The MLOps-specific capabilities are represented by gray boxes.

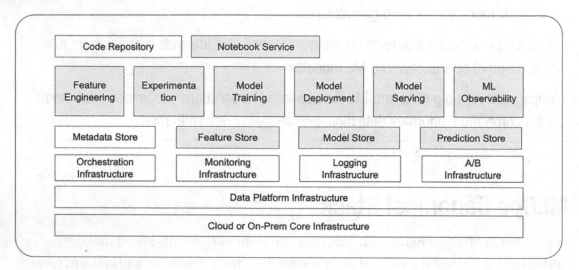

Figure 1-8. *Adapted from the MLOps Blueprint from AI Infrastructure Alliance – https://ai-infrastructure.org/ai-infrastructure-ecosystem-report-of-2022/*

A reader with a keen eye will recognize the MLOps stack is vast and complex, as well as it has many dependencies on the company's foundational infrastructure. One specific infrastructure that MlOps stack has a critical dependency on is the Data Platform infrastructure. Without the ability to store, access, and process data, then it will be almost impossible to successfully operationalize ML projects in any organization. Unless those ML projects are just experimental and small projects.

The MLOps blueprint depicts the core capabilities that are designed to support the ML development lifecycle and the MLOps principles described above.

---

[9] AI Infrastructure Ecosystem of 2022 – https://ai-infrastructure.org/ai-infrastructure-ecosystem-report-of-2022/

MLOps Components

To gain a deeper understanding of the capabilities represented by the components in the MLOps blueprint and how each gray box (component) supports the ML development lifecycle and the principles of MLOps, this section will provide more information about each component.

Feature Engineering

One of the first steps in the ML development lifecycle is about feature engineering, which involves selecting relevant features, creating new features, and scaling and transforming them to make it suitable for the ML algorithm. This step is critical to the success of an ML project, and data scientists often spend a significant amount of time on it.

To accelerate this step, it is helpful to abstract away the underlying complexities and system-related challenges, allowing data scientists to focus on the logic of creating features. In other words, data scientists should be able to define the logic for generating features, and the supporting infrastructure should take care of how to make it happen in an easy, reliable, and scalable manner.

The supporting infrastructure might not be too important for organizations that fall into these buckets:

- In the early part of the journey of applying ML and in the experimentation phase.
- Have only a few ML projects.
- The feature volume is small.

This capability supports the automation principle, and it must for organizations that are looking to scale up the number of ML projects.

Feature Store

Feature store is essentially a centralized repository for storing features and provides feature management, which includes feature registration and feature discovery to understand feature origin, their computation logic, quality, and current status. This enables feature reuse and sharing across many teams that are applying ML.

27

A complete feature store provides storage and access for both offline and online use cases. Online use cases are more challenging due to the low latency access requirements during the online model prediction request path.

For companies that are in the early part of the ML adoption journey, they can start with a simple feature store with limited capability using an S3 bucket.

This capability is a must for scaling AI projects and to operate efficiently with feature reuse and sharing.

The principles this capability supports are continuous versioning, reproducibility, ML training, evaluation, and deployment.

Notebook Service

In the early stages of an ML project, data scientists typically spend a lot of time exploring and experimenting to gain insights into the problem they are trying to solve or analyze the collected data. They may try out different ML algorithms and hyperparameters to establish a baseline model. Notebook services have become popular tools for data scientists, providing a web-based interface for immersive work.

A centralized notebook service with features like access to data and computing resources for model training can greatly enhance data scientist productivity. Open source notebook services like JupyterLab have made it easy for data scientists to launch local instances, but a centralized service can offer additional benefits such as collaboration, version control, and integration with other tools and services.

This capability will only start paying dividends only when organizations are starting to scale their ML projects and investment.

Notebook services support several key principles of MLOps, such as automation, versioning, experiment tracking, and reproducibility. By providing a centralized platform for data scientists to work on, notebook services can help to promote these principles and support a more efficient and effective ML development process.

Model Training

Similar to the feature engineering capability, the goal of this component is to enable data scientists to easily and efficiently train their machine learning models, regardless of their complexity in terms of feature volume, model architecture, and compute resource requirements.

To accomplish those goals, this component should offer the following capabilities:

- The necessary abstraction for data scientists to concisely codify the model training steps in an easy to understand manner.

- The ability to train their models to the needed compute resources, such as CPUs, GPUs, or TPUs.

- Continuous ML training.

- The facility to follow a paved path toward reproducibility and to achieve consistency.

The sections below will dive into more specific details about each of the above capabilities.

Abstraction

Understanding that model training is an interactive process and data scientists are generally well equipped and knowledgeable in selecting the appropriate features, ML model algorithm, and a set of hyperparameters to ultimately produce an optimal ML model for the business problem at hand.

They need an abstraction to express what needs to be done during the model training process in the fewest number of lines of code to

- Split the sample and split the training set

- Specify the ML model algorithm, hyperparameters

- Evaluate and track the model performance metrics

- Access the necessary compute resources to complete the model training in the shortest amount of time

This abstraction is typically provided in a library written in Python and acts as a wrapper on top of other existing libraries as well as to access company infrastructures, such as logging or monitoring.

Technology is constantly evolved. The abstraction eases the migration and pain of upgrading to a new version of the dependent library or switching to a new library that provides better capabilities or accessing newly provided infrastructure.

Compute Resource

As organizations are scaling up their ML projects, undoubtedly their use cases will get more complex and thus might require model training with large amounts of features and using more and more complex ML algorithm architecture.

Access to the needed compute resources at scale to make it possible training a large and complex ML model or reduce the model training time from days to hours or minutes would greatly speed up the iterative model training process and ultimately shorten the time it takes to bring ML models to production.

Launching and managing expensive compute resources whether on cloud or on-prem is typically an engineering task. It is best to require as minimum as possible for data scientists to learn how to do this. Managing the compute resource in terms of launching, shutting down, and cost attribution is best done by model training infrastructure.

Continuous Training

One of the funny software developer moments is "It works on my computer." On a similar note, organizations wouldn't want an ML model deployed to production to be a part of an important business decision workflow, such as approving or rejecting a loan application, if it was trained on a data scientist's laptop.

Ideally, any ML model in production should be trained via a systematic way based on the training code that was coded review and checked into the version control system.

This systematic way is typically engineered as one of the steps in the continuous integration and deployment pipeline that is integrated into the version control system.

This is to ensure the trained ML model was done in a controlled and traceable manner, with audit trail and traceability, and should be fairly reproducible.

Consistency and Reproducibility

As organizations invest more in machine learning and take on more projects, the number of data scientists typically increases, leading to increased collaboration and situations where data scientists may take over projects previously handled by other data scientists who have moved on to other projects or left the company. In these situations, it's important to ensure consistency and reproducibility in the machine learning model training process.

If the model training code is written in a consistent way with high-quality via code review and readily accessible in a version control system for others to see or learn from, this will tremendously boost the data scientist collaboration and productivity across the organization.

The paved path in developing model training code will make it very easy for data scientists to reproduce a previous trained model to iterate on a new version or to debug an unexpected production issue.

Ultimately consistency and reproducibility will increase data scientist productivity and save their sanity.

Model training infrastructure plays a critical role in enabling organizations to scale up their ML investment and projects.

Experimentation

Experimentation is the science part of the data science discipline, and it is an essential part of the ML model training process. Experimentation is about doing the exploration to find the best combination of model architecture, features, and tuning parameters to produce the best performant model to meet the business needs. It requires iterations and tracking of all the experimentation artifacts, similar to conducting a chemistry experiment in chemistry course in college.

This component is not about how to perform experiments, but about enabling all the activities related to and tracking of experimentation.

The essential capabilities this component needs to provide are

- Provide an easy way to send experiment information to a centralized location

- Store the experiment information in a persistent and reliable manner

- Provide an easy way to access the experiment information via APIs or user interface

- Provide an intuitive way to compare two or more experiments so data scientists can easily understand what factors contribute to the change

The principle this component supports is mainly about reproducibility.

Model Store

After an ML model is trained, it needs to be stored in a central repository. From there, it will go through its lifecycle, which includes deployment to various environments such as staging, testing, and production. A central model store serves as a single source of truth for all trained models, facilitating easy access, version control, and collaboration among data scientists and other stakeholders. It also enables organizations to manage the full lifecycle of their models, track their performance, and make informed decisions about which models to deploy and when.

As an organization's machine learning portfolio grows to include a larger number of models, typically around 20 or more, the need for a centralized model store becomes increasingly apparent to overcome challenges with model management and version control.

Model stores bring the following benefits to ML projects:

- Effective management of ML models and their lifecycle

- Model traceability and reproducibility

- Model governance: enable organizations to establish policies and processes for managing models throughout their lifecycle, promoting compliance, and mitigating risks associated with model performance or behavior

- Security: Provide secure storage and access controls for models, protecting them from unauthorized access or tampering

Mode store acts as the bridge between the ML development phase and ML production phase, as depicted in Figure 1-9.

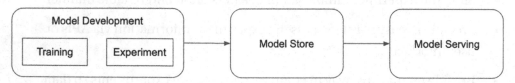

Figure 1-9. *Model store*

Managing ML Models and Lifecycle

Once ML models are officially trained and are candidates for deployment, then they should be stored in and managed by a model store. Acting as a central model repository, it provides visibility, discoverability, auditability, and management of those ML models.

For some companies, the model lifecycle is simple and has only a few stages. For others, due to regulatory compliance, their company's policy dictates a complex model lifecycle. Having a model store manages the defined lifecycle that meets the company's needs is the key to ensuring consistency and meeting compliance.

Model Traceability and Reproducibility

In addition to making it easy to transition an ML model to production by storing the model artifacts and metadata, the model store can help with capturing the audit trail about the deployment information such as who deployed the model, when it was deployed, and any deployment notes.

Traceability is particularly useful and important for companies that would like strong governance and security around model review, model deployment approval, and deployment.

During the model registration process with a model store to get ready for deployment, metadata about the ML model and along with training data information should be captured and stored. If it deems necessary to reproduce a particular model, then all the relevant information is readily available.

The principles this component supports are versioning and reproducibility.

Model Deployment

In the DevOps world, organizations are able to deploy their software automatically and multiple times per day.

In the ML world, similarly, model deployment is a workflow to deploy a trained ML model, hopefully managed by a model store, to production or to rollback a particular deployed model.

The model deployment workflow typically interacts with the model store to retrieve the ML model artifacts, package them in a way that is consumable by the mode serving component, and update model lifecycle status with audit trails.

The second part of the model deployment workflow is to hand over the model artifacts to the model serving component and trigger any necessary changes to enable the newly deployed model to start performing predictions based on incoming requests.

Figure 1-10 depicts the interactions between the model deployment and model store and model serving.

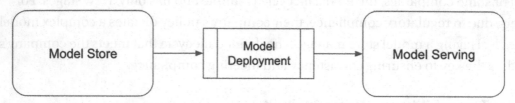

Figure 1-10. *Model deployment*

There are a lot of variations in terms of the low-level technical details and steps in the deployment workflow because different needs and policies various organizations have, as well as the architecture of the model serving component.

The critical part that has direct influence on the speed and frequency of ML model deployment is that the deployment workflow needs to be automated and simple.

The principles this component supports are automation and continuous deployment.

Model Serving

Model serving is where the rubber meets the road, in other words, where the ML trained models are used to generate predictions on production or unseen data. Depending on the particular use case the ML model was trained for, the predictions are generated either during offline, also known as batch, or online. The note below disambiguates batch and online prediction, which typically goes hand in hand with model serving.

Note   Batch prediction vs. online prediction

When the term prediction is being mentioned in a conversation, a common follow-up question is about the context of that prediction. Is it a batch prediction or online prediction?

Batch predictions are generated periodically or on demand and in a batch mode, meaning thousands or millions of predictions are generated together. This approach is well-suited for scenarios where the data to be predicted is available in batches, such as when making predictions based on historical data or when the inputs to the model are collected over a period of time. For example, an e-commerce company might generate batch predictions of customer recommendations based on their past purchase history, generating a set of recommendations for each customer that can be used to inform marketing campaigns. The prediction results are then stored in a storage system such as SQL tables or S3 bucket, and might later be transferred to a fast online system, such as in-memory and distributed storage engines, for consumption to serve online requests. Numerous online recommendation engines use this pattern, such as Netflix movie recommender. Batch predictions are also known as *asynchronous predictions*.

Online predictions are generated based on the incoming on demand requests. Each request typically results in the generation of one to thousands of predictions, but not typically in the millions. The prediction results are sent back to the requestor immediately. That is why online predictions are also known as *synchronous predictions*.

The model serving component is for online prediction use cases and its main responsibility is to host ML models in the form of an online service that might support protocols such as HTTP/HTTPS, REST, or gRPC, and then uses them to generate predictions based on the incoming predictions requests. Essentially the model serving exposes a service endpoint for other services that would like to get the predictions from the hosted ML models.

Model serving marries ML and microservices worlds to enable remote services to leverage ML capabilities in the form of ML predictions. Therefore, the key capabilities of the model serving components is a combination of a typical microservice and ML model needs, such as

- Low latency, scalable, reliable

- ML framework agnostic or the commonly used ones

- Support CPU, GPU, TPU, or other AI-accelerating hardware

- Support model shadowing

- Integrate with A/B experimentation platform

- Support online feature fetching on behalf of the clients

- Support logging of prediction request (features, model id/version, prediction result, etc.)

Model serving component is one of the most critical components in the MLOps stack when organizations start to integrate ML into their online products that service online customer requests, such as recommendation, searching and ranking, fraud detection, language translation, auto-completion, and more.

The principles this component supports are automation and continuous deployment.

Prediction Store

Prediction store is an emerging component that has received less attention and discussion compared to other components. It acts as a central repository for storing the prediction logs for the online prediction use cases. The prediction logs typically contain details such as the input features used to generate predictions, the model generated the predictions, as well as other pieces of metadata and operational metrics.

This information is very valuable for data scientists to

- Debug ML model production issues

- Evaluate the performance of a ML model in shadow mode

- Use as training data for the next version of an existing model

Data scientists need the ability to access, analyze, and process these prediction logs in an easy, efficient, and scalable manner.

The principles this component supports are reproducibility and continuous monitoring.

Note   Josh Tobin, the co-founder of a startup called Gantry, has advocated for a component called Evaluation Store. From his "A Missing Link in ML Infrastructure Stack"[10] presentation, the Evaluation Store is defined as "a central place to store and query online and offline ground truth and approximate model quality metrics."

In addition to supporting the capabilities of the Prediction Store mentioned above, the Evaluation Store also stores the predictions during the training phase and ML model metrics during evaluation phase to enable the ML practitioners deploy ML models more confidently and catch production bugs faster.

ML Observability

ML observability plays an important role in protecting the downside once ML models are deployed into production and integrated into an organization's online products. In other words, to detect and prevent the negative impact to the very business results that the ML models were designed to improve.

More importantly it plays a vital role in ensuring a team doesn't fly blind after deploying their ML models to production and be able to iterate and improve on their models quickly by having the ability to analyze model degradation and to easily perform root cause analysis on model-related issues in production.

ML observability, as a component of MLOps, encompasses several critical aspects, including monitoring, observability, and explainability. These aspects are essential for ensuring that machine learning systems are performing as expected, and for helping teams to diagnose and address issues as they arise.

Monitoring

Monitoring is about trying to answer the questions of what went wrong and when by tracking, measuring, and logging system operational metrics and ML-related metrics, such as accuracy, drift, prediction failures and errors, and more.

---

[10] Josh Tobin, "A Missing Link in the ML Infrastructure Stack", http://josh-tobin.com/assets/pdf/missing_link_in_mlops_infra_031121.pdf

Observability

Observability aims to provide more context or insights into the behavior and performance of ML models, enabling teams to quickly identify and debug issues when they arise. For example, if a model performance starts to degrade, ML teams should be able to easily and quickly determine the root cause, such as changes to production feature data, failures in the feature data pipeline that resulted in stale features, bugs introduced by recent model deployments, or issues with the underlying infrastructure or environment.

Explainability

Explainability tries to help humans understand why a model made a particular prediction or what factors are heavily contributed to such predictions. This information is very helpful to business teams or non-data science teams to gain confidence in the ML model performance and an intuition about how those models behave, as well as to help with finding potential issues while the ML models are in the validation phase or in production.

To address the needs for the above three areas, the key aspects this component needs to provide are

- An easy way to set up the monitoring of various ML-related metrics, such as both model and feature drift, model performance metrics, and along with threshold to trigger alerts for on call ML engineers or data scientists to look into

- An easy way to visualize the various model performance metrics to understand what's going and to analyze the metrics easily so data scientists can pinpoint the issue quickly

- An easy way to see which features play a large role in influencing the prediction outcome

- An easy way to analyze the performance metrics for each ML model

In summary, with ML observability, data scientists will easily and quickly gain insight into model performance issues, will have the ability to understand how ML models make their predictions, and will be able to respond to questions from non-data science teams about the behavior of their ML models.

Note   This component has a big dependency on the availability of the Prediction Store in order to access and compute various aggregations on the prediction log data.

The principles this component supports are automation and continuous monitoring principles.

MLOps Pillars

There are numerous components depicted in Figure 1-8, and their details are described in the previous section. It is helpful to gain a bird's eye view of the MLOps stack by zooming out a bit and logically grouping those components into broader areas, which I called pillars. These pillars capture the key aspects of MLOps stack so they can be easily shared and explained to other teams and to track their progress and maturity while organizations are on their journey of productionizing MLOps. Each of these pillars has sufficient scope and impact, and the boundary is reasonably delineated. Therefore, it makes it very easy to justify that each pillar needs a team to develop and support it.

The four pillars I am advocating for are Feature Engineering, Model Training and Management, Model Serving, and ML Observability, as depicted in Figure 1-11.

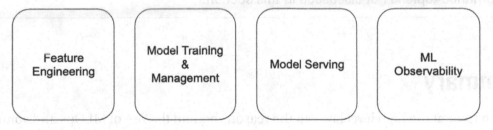

Figure 1-11. *MLOps pillars*

Feature Engineering

This pillar is responsible for all the needed infrastructure to support the activities and processes related to feature generation, feature management, and feature store.

Model Training and Management

This pillar is responsible for all the needed infrastructure to support the activities and processes related to model training, model store, and model lifecycle management.

Model Serving

This pillar is responsible for all the needed infrastructure to support the activities and processes related to model serving and prediction store.

ML Observability

This pillar is responsible for all the needed infrastructure to support the activities and processes related to ML monitoring, observability, and explainability.

Note   ML governance

ML governance comprises a set of policies, processes, and controls designed to manage the lifecycle of ML models, addressing concerns related to ethics, compliance, and risk management. The MLOps pillars are more about the necessary infrastructure to accelerate ML development process. As such, the ML governance topic is not discussed in this section.

Summary

This chapter at the high level lays out the reasons behind the rise of MLOps and some of the challenges in operationalizing ML projects. It touches upon some of the unique aspects of ML projects and how they are different from standard software projects, such as the iterative and experimental natures of ML. Recognizing those differences is the first step in understanding MLOps.

Next, it highlights MLOps as an engineering discipline that combines the best practices and techniques from the other three engineering disciplines: DevOps, Data Engineering, and Machine Learning.

Then, it dissects the MLOps engineering discipline and peels the MLOps onion to gain a deeper understanding by discussing the three main layers: paradigm, engineering discipline, and principles. It is worth repeating that the first and critical step in successful MLOps adoption is the necessary mindset shift toward treating ML artifacts as the first class citizens in the ML project development. Then, by standardizing and practicing the MLOps principles, it will enable organizations to start scaling their ML projects and reaping greater benefits from their ML investment.

Following the discussion about the MLOps engineering practices and principles, it then describes the underpinning technical stack and its components from an engineering's perspective, and not ML's perspective. The canonical stack is meant as a blueprint to provide a scalable and efficient means to automate and scale the development, deployment, management, and monitoring of ML models.

Finally, it proposes a logical way of grouping the components into pillars, such that it is much easier to understand the lay of the MLOps landscape and the scope and boundary of each pillar.

Once an organization understands the best practices, the disciplines, and the canonical stack of MLOps, how should it go about adopting MLOps and integrate it into the large ecosystem? What strategies and approaches should it follow to give it the best chance of success? What are some of the potential challenges that might arise? Chapter 2 will discuss these important questions and share some tips that might be helpful to organizations that are looking to adopt MLOps or facing challenges while adopting MLOps.

CHAPTER 2

MLOps Adoption Strategies and Case Studies

There is no doubt that we are in the heady days of AI/ML operationalization. According to a Gartner article about the "IT Budgets Are Growing. Here's Where the Money's Going,"[1] AI/ML is at the top of the list of technologies, and nearly half of CIOs say they now employ AI/ML or intend to within the next 12 months. The "AI Adoption in the Enterprise 2020"[2] report from O'reily confirms such statistics and further shared that AI/ML adoption is pervasive across many industries, such as financial services, education, healthcare, manufacturing, retail, and more.

Maximizing the ROI from AI/ML investments and shortening the time to value are critical goals for organizations. One key factor in achieving those goals is effective implementation of MLOps. While having a solid understanding of MLOps is a good starting point, translating that understanding into a successful implementation can be challenging.

Similar to any technology adoption, successful MLOps adoption requires a methodical approach and a set of strategies to manage the MLOps-specific nuances, as well as numerous factors that need to be taken into consideration.

This chapter will start with outlining a high-level strategy and highlight a few important dimensions organizations need to consider to ensure a successful MLOps adoption. After that, it will offer a couple of approaches to making MLOps a reality and will share some of the pros and cons of each approach.

---

[1] IT Budgets Are Growing. Here's Where the Money's Going – www.gartner.com/en/articles/it-budgets-are-growing-here-s-where-the-money-s-going

[2] AI Adoption in the Enterprise 2022 – www.oreilly.com/radar/ai-adoption-in-the-enterprise-2022/

© Hien Luu, Max Pumperla and Zhe Zhang 2024
H. Luu et al., *MLOps with Ray*, https://doi.org/10.1007/979-8-8600-0376-5_2

For those organizations that are considering adopting or in the early stage of adopting AI/ML, the hope here is that some of the details in this chapter will be helpful in either adjusting their current adoption approach or enabling them to start their journey on solid ground.

Numerous large and successful companies have gone through the journey of building out their MLOps infrastructure, especially the ones that were born during the Internet era or shortly after. The last part of this chapter will go over a few successful MLOps adoption case studies.

Let the journey begin!

Adoption Strategies

According to the "AI Infrastructure Ecosystem 2022"[3] report, only 26% of their survey respondents are very happy with the current state of their AI/ML infrastructure. As an industry, it looks like there is a lot of room for improvement. This particular survey response speaks to the fact that figuring out and putting together the right AI/ML infrastructure that fits the specific needs of an organization is a very challenging task. However, this same survey pointed out the good news that the majority of organizations were able to swiftly reap the rewards from their AI/ML infrastructure investment in two years or less.

Organizations that are adopting AI/ML will need to have an MLOps infrastructure, which is also known as AI/ML infrastructure or ML platform. Let's quickly remind ourselves that MLOps's goal is to help organizations to operationalize ML consistently, effectively, and efficiently so that they can benefit from AI/ML's power to increase their competitive advantage, top line and/or bottom line, and more importantly will get the ROI from their investment in the ML projects.

Similar to previous technological advancements, there isn't a single overarching technology adoption strategy that would address the uniqueness and the various needs of each company. In addition, the success criteria of MLOps adoption on one organization are very likely to be different than the next one.

---

[3] AI Infrastructure Ecosystem of 2022 – https://ai-infrastructure.org/ai-infrastructure-ecosystem-report-of-2022/

Before deciding on an approach for building their MLOps infrastructure, an organization or the leadership team should figure out these two critical aspects:

- Alignment between business goals with MLOps infrastructure goals
- Assessment of specific MLOps needs

Goals Alignment

The alignment between business goals and MLOps infrastructure goals is an obvious one; however, it can't be taken lightly. It is important to remember that MLOps infrastructure is a means to achieve the ultimate goal of harnessing the power of AI/ML to add value to their business goals and to get the ROI in the ML project investment.

The benefits of alignment include

- Knowing the business goals and their priorities will help influence the sequence of putting together your MLOps infrastructure and recognize which areas deserve more attention

- The MLOps infrastructure adoption discussions will be much easier since everyone is aligned on the business goals and their priorities

- Securing funding for talent and buying vendor solutions will be easier to justify with a clear understanding of the ROI from the business goals

It is understandable that business goals will evolve over time due to various factors, and MLOps infrastructure goals must adapt accordingly. If the MLOps infrastructure goals do not keep pace with the changing business needs, there is a risk that the infrastructure will be perceived as a cost center rather than a valuable asset.

MLOps Need Assessment

The MLOps canonical stack described in Chapter 1 provides a comprehensive overview of the various essential components that a typical MLOps infrastructure requires. While this stack serves as a useful blueprint for understanding the various components involved, simply implementing each component in sequence may not be sufficient to achieve a successful outcome. There are several key dimensions that must be carefully considered.

Organizations differ in a variety of ways, including their business domain, size, the current state and maturity of their technology that are related to or supporting ML projects, team size and talent in Data Science and MLOps infrastructure, and most importantly company culture. These factors influence an organization's unique needs and capabilities in terms of MLOps infrastructure and implementation.

The following sections will dive deeper into each of these areas, as well as highlight how they might impact the sequence of putting together an MLOps infrastructure, the level of emphasis each component needs, and additional considerations.

Use Cases

AI/ML is quite a powerful tool for businesses due its versatility of being able to help with solving many types of decision-making-related business problems.

The various ML use case types an organization needs to tackle is mainly driven by the business domain that they are in and the use case volume is primarily influenced by their company size and the number of business functions.

Table 2-1 contains a set of well-known ML use cases across a few business domains.

Table 2-1. *Business domain and examples of ML use case*

Business Domain	ML Use Cases
Healthcare	Medical diagnosis, patient monitoring
Marketing	Personalized advertising, sentiment analysis
Security	Facial recognition, intrusion detection
Transportation	Route optimization, autonomous vehicles
Finance	Loan approval, fraud detection, credit scoring
Telecommunications	Customer churn prediction, targeted marketing, network optimization
Energy	Predictive maintenance, demand forecasting
Social Media	Personalized recommendation, advertising optimization
Retail	Demand forecasting, price optimization

Let's pick a few use cases to discuss their specific needs and how that might influence the approach of putting together an MLOps infrastructure.

Fraud Detection

The first use case to discuss is fraud detection from the Finance business domain, which is something that can be easily understood, especially for credit card holders. In an ideal world, the finance organizations would like to prevent fraudulent activities before they happen, but that is not always feasible. The next best hope is to quickly detect those activities shortly after they appear so those bad activities can be mitigated and to minimize the damage.

To achieve this goal, it requires the ability to perform online/real-time predictions or inference and likely at scale. From the MLOps infrastructure perspective, it needs to provide a scalable, low latency, and highly reliable prediction service. This doesn't mean the other parts of the MLOps stack are not important, rather this use case suggests that the online prediction service deserves higher priority as a part of the strategy of putting together an MLOps infrastructure.

Churn Prediction

Another interesting use case to discuss is the churn prediction from the Telecommunications business domain, which is a highly competitive market. Churn prediction use case requires a comprehensive understanding of the customers, their profile, tendency, activities, and more. This ML use case usually requires the customer data from many data sources, and this means the chance of something going wrong related to the data is pretty large. For any midsize to large telecommunication company, likely there will be multiple dozens of ML models. The weekly and monthly customer churn rate is highly monitored and tracked by business leaders.

For this use case, it is quite important to easily monitor and debug model performance degradation.

From the MLOps infrastructure perspective, along with automation of the ML pipelines, it needs to provide fairly comprehensive ML observability capabilities to enable data scientist and business domain experts to perform ML model tracing and easily visualizing the drift to understand what's going as well as to compare model performance between the baseline and production.

Loan Approval and Credit Scoring

In highly regulated industries, such as healthcare, finance, and transportation, compliance with relevant regulations is of paramount importance. This underscores the importance of robust ML governance in these industries, which involves implementing processes and mechanisms to ensure that ML systems are fair, transparent, and accountable.

For use cases like loan approval and credit scoring, any perceived demonstration of bias against a protected category could trigger investigations from a governmental agency.

MLOps, which involves the development and deployment of ML models in a systematic and repeatable manner, is a critical component of ML governance.

Note   ML governance

ML model governance isn't a piece of technology, but rather it is a process that is borrowed from the corporate governance and tailored to ML models. The general principles are still applicable, such as access control and activity tracking.

According to the article "What Is Model Governance?"[4] from DataRobot, ML model governance is the overall process for how an organization controls access, implements policy, and tracks activity for models and their results.

The specific capabilities the MLOps infrastructure needs to provide to meet the ML model governance might include the access control management for all models in production, tracking model history, monitoring models and their predictions, and more.

The extracted insights from understanding the organization's core ML use cases and their priority and then working backward from their needs will help with determining which parts of the MLOps infrastructure are more important than others and a reasonable starting point in building out the MLOps canonical stack.

---

[4]What is Model Governance from DataRobot, 2020, www.datarobot.com/blog/what-is-model-governance/

Technology

MLOps infrastructure is just one of the pieces of the puzzle of an effective ML development ecosystem. The other two important infrastructure pieces are DevOps and DataOps. DevOps infrastructure is the automated and streamlined set of tools and processes facilitating collaborative software development and deployment, providing a foundation for efficient ML development. DataOps infrastructure streamlines collaborative data management through automated tools across the entire data lifecycle, which feature engineering, which is particularly relevant to feature engineering in machine learning workflows. The interplay between these three infrastructures is depicted in Figure 2-1.

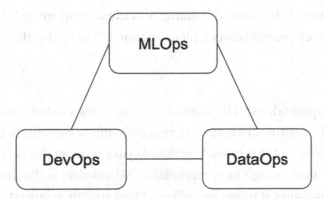

Figure 2-1. *The interplay between MLOps, DevOps, and DataOps*

The call to action is here to assess the maturity of the current DevOps and DataOps, highlight the MLOps dependency on them, and advocate for areas of improvement if the identified gaps have measurable impact on the ML development lifecycle.

Two other important infrastructures that have a large influence on the effective ML development ecosystem are compute and experimentation infrastructure.

The compute infrastructure plays a key enabler for organizations that are looking to scale their ML projects in terms of generating large volumes of features, training large and complex models, and scalable model inference. It offers several specific capabilities tailored to the unique demands of machine learning workflows. Its scalability ensures dynamic resource allocation, optimizing performance, and responsiveness. GPU acceleration supports faster model training, particularly for complex deep learning models. Containerization and orchestration tools enable consistent packaging and deployment of ML models, while parallel processing enhances efficiency in tasks like hyperparameter tuning. Compute infrastructure facilitates efficient model serving and resource monitoring, enabling cost-effectiveness and optimal resource utilization.

For the ML models that will eventually be integrated into an organization's online products and services, the speed of iterating on those models will lead to a faster time to deploy them in production. While iterating on the ML models to figure out the best performing one, it requires the ability to experiment, test, and validate the model performance in production via A/B testing. As a result, the experimentation infrastructure that enables quick and easy ways to run A/B testing and provides easy and intuitive analysis of experiment results is crucial to the success of MLOps infrastructure and the over ML project investment.

The call to action here is to assess the maturity of these supporting infrastructures, identify gaps or ML development velocity blockers, and advocate for them.

People

Successful MLOps adoption doesn't happen in a vacuum. It's a collaborative effort involving the MLOps infrastructure team, internal customers, stakeholders, and key decision-makers. Without this cross-functional synergy, even the best MLOps infrastructure wouldn't be enough to operationalize ML models at the desired pace.

For smaller organizations, it might be sufficient to start with a virtual team composed of team members from various teams, such as data engineering, devops, and data science team. However, as the need to deploy and scale ML models in production increases, particularly for integrating them into the core online products and services to deliver business value, a dedicated MLOps team may become necessary. This team should consist of talented individuals who can develop, maintain, and support MLOps infrastructure for the rest of the organization, as well as establish and enforce standards and best practices.

Identifying and aligning internal customers, stakeholders, and key decision-makers is a crucial component of successful MLOps adoption. The first step is to clearly understand who the customers and stakeholders of the MLOps infrastructure are. The next step is to establish ongoing two-way communication with these individuals, involving them in the MLOp infrastructure strategy and roadmap, soliciting their feedback and requirements, and more importantly, anticipate their needs.

Culture

It is fair to say most organizations have a decent understanding about the value AI/ML can bring to their business, even for the ones that aren't born digital. For those organizations where AI/ML projects face formidable culture and organizational barriers, it is important there are active plans coming from top to address them and lower them.

The discussions about the cultural challenges in this section are made with the assumption that AI/ML adoption is one of the top initiatives at an organization.

Company culture refers to the beliefs and behaviors that guide how a company's employees and management interacts, and helps define its identity. Successful MLOps adoption needs to consider the culture element as a part of the equation, especially around these dimensions: risk tolerance, execution velocity, decision-making process, and collaboration style.

Risk Tolerance

Undoubtedly during the journey of putting together an MLOps infrastructure, there will be situations that involve risk-taking, such as evaluating and adopting commercial vendor solutions, exploring new open source technologies, and more.

It is crucial to have a good sense about the risk tolerance level of the organization culture, whether that is at the conservative or moderate level. This will influence the amount of effort and time it would take to control the risk, as well as the amount of communication to the broader audience during the risk assessment and alignment phase.

For example, a relatively new piece of open source technology has been identified and proven to be effective at improving model training velocity. Since that piece of technology is relatively new, therefore, there will be unknown risks in handling certain scenarios. An organization's risk tolerance level will have a large influence in getting a green light to the adoption. The MLOps infrastructure team will need to consider the appropriate amount of time to invest in vetting the technology and to convince others.

Execution Velocity

An organization's execution velocity is usually associated with the current state of its lifecycle. Small organizations in its early lifecycle stage usually need to move fast at execution to either complete a proof of concept of their product idea or to achieve a product-market fit.

The organization's execution velocity will have a large influence on the pace of MLOps infrastructure adoption as well as the level of expectation how quickly the MLOps infrastructure needs to demonstrate impact to the rest of the organization. For a fast moving organization, it is important that the MLOps infrastructure team needs to prioritize and demonstrate incremental progress at a steady pace over perfection.

Ultimately the MLOps infrastructure team needs to operate at the pace that their customers and stakeholders operate at in order to contribute to the organization's AI/ML goals and make them happy.

Decision-Making Process

Most successful organizations have a fairly healthy decision-making process for important decisions, whether that is a consultative or collaborative style.

Undoubtedly during the putting together of an MLOps infrastructure, there will be many important decisions that will need to be made, such as technology and vendor solution adoption, the model deployment process, the level of rigor around model access control, and more.

Understanding the organization's decision-making process will help the MLOps infrastructure team to know how much time and effort is required in driving particular decisions, as well as involving the right decision-makers and influencers.

Collaboration Style

ML projects often involve multiple stakeholders, including data engineers, data scientists, ML engineers, and MLOps infrastructure team. Effective collaboration between these personas is critical for the successful adoption of MLOps. If collaboration is not in a healthy state, the MLOps infrastructure team should prioritize strengthening it. This may require allocating additional time and resources to proactive collaboration efforts, such as establishing clear communication channels, defining roles and responsibilities, and fostering a culture of collaboration and trust.

By prioritizing healthy collaboration, the MLOps infrastructure team can help ensure all stakeholders are working toward a common goal and that the MLOps adoption process is smooth and effective.

A sizable qualitative empirical study[5] about the socio-technical aspects of building ML-enabled software identifies three main areas in the organization of ML projects causing problems: leadership vacuum at management level, organization silos, and communication within an organization. Some of these aspects are heavily influenced by the culture of an organization; therefore, an MLOps infrastructure team needs to be aware of such challenges and take proactive steps to overcome or minimize them.

Maturity Level

Similar to other technology adoptions, MLOps adoption is not a one-time event, but rather it is a journey with continuous improvement to meet the ever-changing needs of organizations that are progressing from exploring and prototyping AI/ML projects in a small number of business functions to building advanced AI solutions across many business functions.

The MLOps maturity levels are used as a means to track the progress and sophistication of an MLOps infrastructure. It is best used as an assessment tool to understand the current status of the MLOps infrastructure at an organization, to know what the next step will be, and how close or far away the advanced maturity level is.

As an industry, there is a consensus on establishing an MLOps maturity model; however, there is less agreement on the standard set of maturity levels. The two well-known and commonly referenced maturity models are from Google Cloud[6] and Microsoft Azure[7], as depicted in Figure 2-2 and Figure 2-3, respectively. The specific details about each of the levels of these two maturity models can be easily found with a quick search on the Internet.

---

[5] Alina Mailach, Norbert Siegmund, "Socio-Technical Anti-Patterns in Building ML-Enabled Software" https://sws.informatik.uni-leipzig.de/wp-content/uploads/2023/01/socio-technical-anti-patterns-icse2023.pdf

[6] Google Cloud Maturity Model – https://cloud.google.com/architecture/mlops-continuous-delivery-and-automation-pipelines-in-machine-learning

[7] Microsoft Azure Maturity Model – https://learn.microsoft.com/en-us/azure/architecture/example-scenario/mlops/mlops-maturity-model

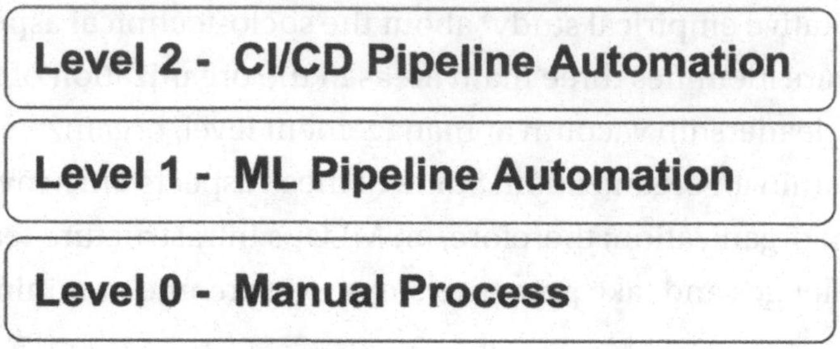

| Level 2 - CI/CD Pipeline Automation |

| Level 1 - ML Pipeline Automation |

| Level 0 - Manual Process |

Figure 2-2. *MLOps maturity model from Google Cloud*

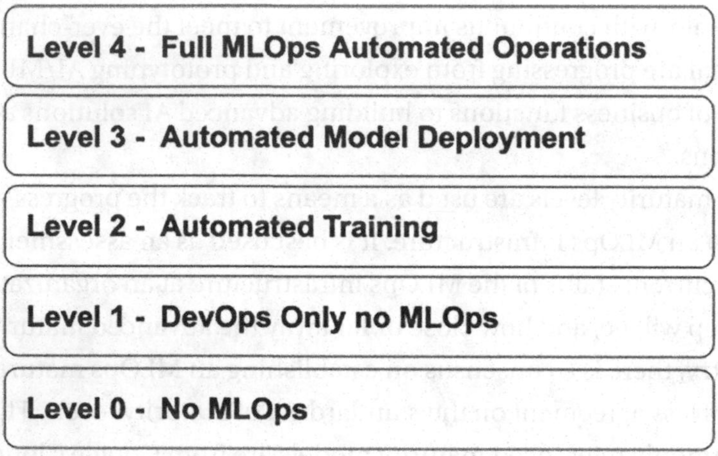

| Level 4 - Full MLOps Automated Operations |

| Level 3 - Automated Model Deployment |

| Level 2 - Automated Training |

| Level 1 - DevOps Only no MLOps |

| Level 0 - No MLOps |

Figure 2-3. *MLOps maturity model from Microsoft Azure*

The common theme across these two maturity models is about the advocacy for automation, which directly increases the velocity from model development to model deployment to production while ensuring proper version control and reproducibility.

It is good to adopt a particular maturity model to follow; however, don't let it keep you up at night wondering which one is the right one. Deciding on which level an organization should be at and how fast to advance from one level to another requires a good understanding of the specific needs of their ML use cases and the maturity of the various underlying supported infrastructures and components in the MLOps canonical stack.

It is fair to say that achieving full automation or reaching the highest level will enable organizations to harness the power of AI/ML to create business value quicker and maximize the chance of seeing the return on their AI/ML investment.

Instead of focusing on the automation level, Josh Poduska, the chief data scientist at Domino Data lab, offers a maturity model based on the MLOps capabilities and business value in his "The Seven Stages of MLOps Maturity" blog[8]. Readers can get more details about each maturity level and the associated value on Josh's blog.

The last section of his blog suggests each organization to assess their own MLOps journey to understand where they are at on the maturity curve and make plans to advance to the next level. It then highlights the key to going beyond the inflection point in value is to tightly integrate all the capabilities in a data science-centric manner.

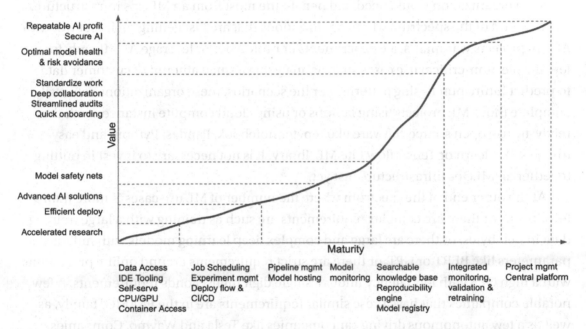

Figure 2-4. *Enterprise MLOps maturity curve. Source: Adapted from an image by Josh Poduska (https://towardsdatascience.com/the-seven-stages-of-mlops-maturity-ccb029530f2)*

The further an organization is able to advance to the higher maturity level, the more their MLOps infrastructure can help with maximizing the efficiency, productivity, and output of their data science teams, which leads to a strong return on their AI/ML investment.

---

[8] "The Seven Stages of MLOps Maturity" https://towardsdatascience.com/the-seven-stages-of-mlops-maturity-ccb029530f2

MLOps Infrastructure Approaches

When organizations are looking to adopt technologies or infrastructure, such as data infrastructure, IT system, and DevOps infrastructure, they typically consider these three approaches: build your own, buy an end-to-end solution, best of breed. The first two cover the extremes and the last one covers the middle ground. These approaches are also applicable for putting together an MLOps infrastructure.

Before discussing the pros and cons of each of these approaches, let's first establish when an organization would need and benefit the most from an MLOps infrastructure.

One end of the spectrum where organizations that are just getting their feet wet with AI/ML projects via small-scale experiments or prototypes or leverage AI/ML only for a few simple, non-critical use cases such as analyzing a small amount of customer data to predict future purchasing patterns. For the scenarios, these organizations can easily complete those ML projects using laptops or using cloud compute instances with the ready-to-use open source software like Jupyter notebook, Pandas, Python, and easy-to-use scikit-learn or Tensorflow Lite ML library. It is not necessary to invest in putting together an MLOps infrastructure just yet.

At the other end of the spectrum where the number of ML use cases is more than a few dozens or there are complex requirements are such as training with a large training data in petabytes, or there are large and complex deep learning models with millions of parameters like BERT or GPT, or there are strict requirements around online predictions with a high QPS with low latency under a single digit milliseconds requirements. A few notable companies that have these similar requirements are in the FAAMG[9] family, as well as a few autonomous driving car companies like Tesla and Waymo. Companies that fall into the right side of the spectrum tend to be pretty good at operationalizing ML models because they are the early adopters, and have financial and talent resources. Very likely these companies invest resources, and talent in building their own in-house MLOps infrastructure, which consists of a mix of open source tools, internally developed tools, and potentially a few vendor solutions to meet their specific and unique needs.

---

[9] FAAMG is an abbreviation for Meta (formerly Facebook), Amazon, Apple, Microsoft, and Google.

Note   At the time of writing this chapter, ChatGPT from OpenAI is dominating the AI/ML news cycle. ChatGPT stands for Chat Generative Pre-trained Transformer, and it was created and launched by OpenAI in November 2022. In layman's terms, it is a fairly intelligent chatbot that was trained and designed to hold natural conversations.

The interaction with ChatGPT is mainly through a dialogue format with follow-up questions. Users submit a wide range of questions to ask ChatGPT to write a short report, summarize a paragraph, generate a lesson plan for a particular topic, write code to develop a website, devise a short travel plan or a set of sightseeings, and much more.

More details about ChatGPT can be found at `https://openai.com/blog/chatgpt`.

There are a large number of companies that fall into the middle region of the spectrum, and they might or might not put together their own MLOps infrastructure, depending on where they are on their journey of adopting ML and their level of ML proficiency, which comes with time and experience. The term "Reasonable Scale"[10] was mentioned in the "ML and MLOps at a Reasonable Scale" blog to represent this group of companies. The oval shape in Figure 2-5 represents "Reasonable Scale" companies, which could be late adopters with a low ML proficiency and a reasonable number of ML use cases or digital native AI startups with a reasonable ML proficiency and a small number of ML use cases.

---

[10] Ciro Greco, ML and MLOps at a Reasonable Scale, 2021 `https://towardsdatascience.com/ml-and-mlops-at-a-reasonable-scale-31d2c0782d9c`

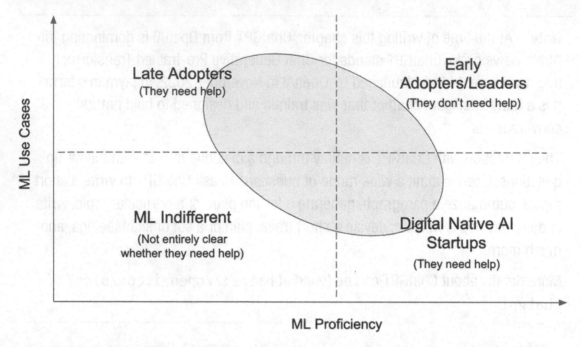

Figure 2-5. *Reasonable scale companies. Source: Adapted from an image by Ciro Greco (https://towardsdatascience.com/ml-and-mlops-at-a-reasonable-scale-31d2c0782d9c)*

As these companies scale up ML investments and projects as well as gain their ML proficiency, it is highly probable they will consider putting together an MLOps infrastructure with a certain amount of tools from cloud providers, specialized commercial vendor solutions, and open source community.

The approach an organization needs to consider should not be a one-time event; smart organizations typically revisit their decisions once a while to see if any necessary adjustments are needed.

One important wrinkle that plays a significant role in deciding the approach is the maturity of the MLOps space. It is still in the early days, and there are numerous innovations and emerging technologies from the open source community that are introduced at a rapid pace. On the vendor solution side, there is a sizable array of options to choose from and they are being developed, enhanced, and evolved just as fast. There are reasonably clear market leaders for certain parts of the MLOps stack, and for others there is a set of potential options.

The following sections will highlight a few important notes and share a few suggestions for each of the options. They are meant for companies that are at the beginning of MLOps adoption journey as well as for those that are in the middle of it.

Build

This approach refers to building the entire MLOps infrastructure (MLOps canonical stack) from the ground up. The complexity is quite high, and it will require a large investment in engineering resources and talent. In addition, the long-term maintenance and tech-debts will also need to be considered as a part of the evaluation and decision process.

The high-tech and early adopters of AI/ML, such as Google, Netflix, Meta, Uber, LinkedIn, Tesla, and more, are the ones that built their own MLOps infrastructure from scratch, mostly due to the combination of necessity and their unique needs and requirements.

In the early days about 5 or more years ago, there were no available commercial vendor solutions and very limited open source technologies in the MLOps infrastructure space. Out of necessity, they would build out their own infrastructure and invent along the way to meet their needs.

Most of these companies have a sizable ML use case and span across many teams, and therefore the investment in building a centralized MLOps infrastructure is an easier case to make and justified. Built once and leveraged by multiple teams is fairly easily understood by the leadership time.

Another aspect to justify the need to build is their unique requirements, and one of them is operating at scale due to their massive customer base in hundreds of millions.

In recent years, some of these companies have started shifting their strategy to leverage some of mature and robust open source technologies and commercial products to complement their infrastructure to either leverage new capabilities or reduce maintenance cost. Examples of these cases include experiment tracking and Ml observability.

According a survey from AI Infrastructure report,[11] only 20% of companies built their infrastructure in-house. It is not unreasonable to forecast that this percentage will go down over time as more and more innovations emerge from both the open source communities and MLOps commercial vendors.

---

[11] AI Infrastructure Ecosystem of 2022 – https://ai-infrastructure.org/ai-infrastructure-ecosystem-report-of-2022/

For companies in the Reasonable Scale spectrum, it is highly recommended to stay away from this approach, unless they have some very unique requirements or special needs, such as strict governance or compliance or ML becomes an integral part of their competitive advantage and their engineering team is getting more advanced and proficient.

Buy

This approach refers to buying and adopting a single, unified end-to-end MLOps infrastructure to all the needs of an organization. The built approach described above is at one end of the spectrum, and this one is at the opposite end.

This approach seems to be an attractive one for certain organizations based on their operating style, how far along they are in their ML/AI adoption journey, the number of use cases, and more. However, there are few important aspects to keep an eye out for.

This approach is a good option to consider for organizations that fall into the following areas:

- Early in their ML/AI journey and experimenting or working a few ML/AI POCs. Leveraging an end-to-end MLOps infrastructure from cloud vendors would greatly speed up their goals and gaining ML proficiency.

- Digital native startups with a small team, limited bandwidth, and resources. Leveraging an end-to-end MLOps infrastructure from cloud vendors would enable them to focus on the ML use cases and not be bottlenecked by building infrastructure.

- The core competitive advantages of their business. Organizations that are in real estate, retail, education, construction, and similar, MLOps infrastructure are most likely not their main focus, and it makes sense to lean more on managed services.

A few important considerations organizations need to keep in mind before deciding to adopt an end-to-end MLOps infrastructure: MLOps maturity, breadth-vs-depth.

For MLOps, it is still in the early days and there are numerous innovations, and new and emerging technologies are introduced at a rapid pace. According to Gartner Hype Cycle for Data Science Machine Learning, 2022, MLOps is in the "Peak of inflated Expectations."

At time of this writing, MLOps is at the tail end of the second phase of the hype cycle. In the context of an end-to-end MLOps infrastructure, there isn't a single solution or all-in-one platform that meets every need in the ML development lifecycle that includes building, training, deploying, and monitoring ML models.

There isn't a lack of attempts though. Several notable cloud vendors, such as Amazon AWS, Google Cloud, and Microsoft Azure, are positioning themselves as providers of such all-in-one platforms, and they are rapidly expanding capabilities to their platform over time. With time, undoubtedly these platforms will become highly developed and be able to cover a broad range of needs for organizations to pick and choose from.

Until then, there is a tendency for these platforms to expand their footprint in terms of capabilities, but the specific features of each major component is a bit shallow. In other words, the current focus is on breadth, not depth, which isn't a bad strategy to meet their customer evaluation checkboxes. Examples of the lack of depth include highly scalable model serving engines and state-of-the-art monitoring, explainability, and observability.

Different ML use case types have slightly different needs. ML use cases where the training data are videos, images, or sounds have slightly different needs than the ones with structured data, for example, labeling or annotation. If the majority of the ML use cases of an organization are in the not the structured data, then it is worth double checking how well an end-to-end MLOps infrastructure can support the needs of those use cases.

Until there are clear leaders in the end-to-end MLOps infrastructure and the MLOps maturity is further along in the hype cycle, it is best not to expect a complete, unified, and end-to-end MLOps infrastructure that can meet all use cases, both in breadth and depth.

Hybrid

This approach refers to buying solutions for some part of the MLOps infrastructure, and for the remaining portions, either build them in-house or leverage open source solutions.

As organizations in the Reasonable Scale area move through their AI/ML adoption journey, their needs will change and evolve with time due to the type of ML use cases they need to tackle, the number of ML use cases they need to support and scale, and their ML proficiency advances. The hybrid approach gives these organizations the most flexibility in putting together an MLOps infrastructure. However, flexibility comes with additional responsibilities.

The flexibility this approach provides to organizations is in the form of modularity and freedom.

For those parts of the MLOps infrastructure that are common or don't vary that much across major types of use cases, organizations might want to rely on one or two vendor solutions. Examples of the parts in discussion are data processing, pipeline orchestration, versioning and lineage, and experiment tracking. It is reasonable to think of these parts as the hub of the MLOps infrastructure.

For the other parts, organizations can rely on vendor solutions or open source solutions. For the specific needs, it is reasonable to go with vendor solutions provided they are the leaders into those areas and can satisfy those needs. Examples of these specific needs include synthetic data generation, highly customized model governance, and advanced capabilities in ML observability and explainability.

On the other side of the coin, the provided flexibility this approach enables demands some effort in integrating these specialized solutions together with the hub. The burden can be lessened if those solutions have clean, well-documented APIs and integration points.

Two other aspects that come with the flexibility are cost and support. Undoubtedly there will be an associated cost with buying and integrating multiple solutions, and this will need to be weighted against the other two approaches. The amount of cost will be a function of the number of specialized solutions that need to be integrated into the hub and the complexity.

The second aspect is about support. With multiple vendor solutions, then the "one throat to choke" leverage is not available, and organizations will need to exert additional effort in managing the overhead.

Putting together an MLOps infrastructure is no doubt a challenging endeavor due to the specific needs organizations have, their preferred style of operating infrastructure, their use case types and volume, their ML proficiency, and more. Among the three approaches, the hybrid one, which provides modularity and flexibility, can be the most effective one to build and evolve an MLOps infrastructure to meet current and future needs.

Before moving on to describing a few notable MLOps infrastructure case studies, let's summarize the key points in this section.

- It might be obvious, but it is important and worth repeating. Regardless of which option an organization decides to go with, it is highly recommended to start with a good understanding and an inventory of the ML use cases that need to be tackled now and in the near future across the various functions.

- Ensure to incorporate the recognition of the current MLOps maturity level and understanding there will be better and innovative solutions in the future into your MLOps infrastructure decision-making process.

- Treat the suggestions mentioned in this section as rule-of-thumb, not gospel truth.

Let's end this section with a relevant tweet about build vs. buy from an ex-CTO of Better.com, Erik Bernhardsson, "It's interesting how one of the most important jobs of a CTO is vendor/product selection and the importance of this keeps going up rapidly every year since the tools/infra space grows so fast."[12]

MLOps Landscape

Without any doubts, the MLOps landscape has been exploding over the last 7 years. Multiple billions of dollars have been invested into numerous MLOps-related companies.[13] There is a glut of innovations from the commercial vendors and open source communities. The academia has taken an interest in MLOps and formalizes a community around it that researches on the intersection of Systems and ML and hence a conference,[14] MLSys to take this forward.

Any attempts to capture the MLOps explosion will be shortly out of date, but useful to have a sense about the magnitude. The "Machine Learning tools Landscape v2"[15] blog by Chip Huyen mentioned there are about 300 MLOps tools as of December 2020 and provided a nice visualization to explore the various categories. There are numerous innovations in the ML observability area of MLOps. "The State of MLOps" website collected this evidence and shared it in an Airtable,[16] which indicated there are about 50 companies that are focusing and providing solutions in this particular area.

---

[12] Erik Bernhardsson, his tweet on Twitter (September 2021, https://twitter.com/bernhardsson/status/1443202575466180617)

[13] The State of MLOps, Dr. Ori Cohen's Research (www.stateofmlops.com/)

[14] MLSys: The New Frontier of Machine Learning Systems (https://arxiv.org/abs/1904.03257)

[15] Chip Huyen, Machine Learning Tools Landscape v2, https://huyenchip.com/2020/12/30/mlops-v2.html

[16] The State of MLOps, Dr. Ori Cohen's Research (www.stateofmlops.com/)

As a whole, the ML practitioner community has grown a lot in terms of being more educated via blogs, books, conferences, and more savvy from leveraging some of the shared best practices.

For any organizations that are looking to put together an MLOps infrastructure or to assess the effectiveness of their current MLOps infrastructure, it is beneficial to have a high-level understanding of the current landscape, who are the movers and shakers.

Platforms and Tools

Regardless of which approach an organization is going to pick, it is good to clearly understand the distinction between end-to-end platforms and specialist tools.

- Specialist tools are meant to support a particular part of or subset of the ML development lifecycle.

  - Examples of these specialist tools are Arize for monitoring and observability and MLFlow and MetaFlow for ML project frameworks.

- An end-to-end platform is meant to support the entire ML development lifecycle – feature development, model development and training, modeling serving, and model monitoring. Essentially an end-to-end platform consists of a collection of specialist tools that have been designed to work together.

  - Examples of end-to-end platforms are AWS SageMaker, Google Vertex AI, and Microsoft Azure.

At the high level, end-to-end platforms and specialist tools fall on the spectrum of generality and specialization, as depicted in Figure 2-6.

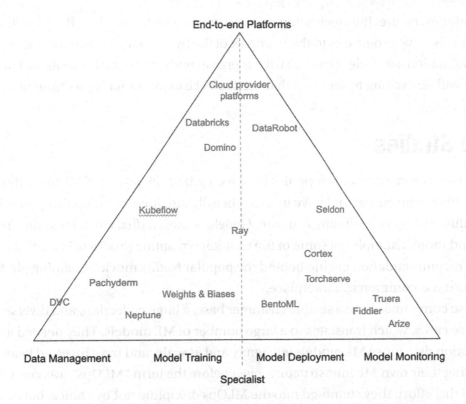

Figure 2-6. *End-to-end platform and specialist tools Spectrum. Source: Adapted from an image in Guide to Evaluating MLOps Platform from Thoughtworks[17]*

The specialist tools on the bottom-left are designed to support feature development and model training. The ones on the bottom-right are designed to support model deployment and monitoring. The ones closer to the top of the triangle, such as cloud provider platforms, fall into the end-to-end platforms.

Some specialist tools start out with fulfilling only one area of ML development, but over time, its footprint expands to support the adjacent areas. In other words, they specialize in more than one area.

On the platform side, the ones from cloud vendors, such as AWS, Google Cloud, and Microsoft Azure, seem to be the front runners in terms of the capabilities; a few other platforms are not too far behind. A great resource to understand the differences among the platforms is the MLOps Platform Comparison matrix[18] provided by ThoughWorks.

---

[17] Thoughtworks, "Guide to Evaluating MLOps Platform (November 2021)," `www.thoughtworks.com/en-us/what-we-do/data-and-ai/cd4ml/guide-to-evaluating-mlops-platforms`

Similar to the previous technological waves that came before the MLOps technology wave, as this wave progresses to the later part of the hype cycle, undoubtedly there will be consolidations, and clear leaders will emerge in both the specialist tools and platform areas. It will be exciting to see what that will look like in five or ten years from now.

Case Studies

Most of the largest Internet companies have recognized the power of ML/AI early on since 2010 or even earlier and have invested heavily into incorporating this power in their online products, including Amazon, Google, Meta, Netflix, Uber, LinkedIn, Twitter, Tesla, and more. Examples of some of the well-known online products include the famous recommendation engine behind the popular Netflix movie streaming platform and Amazon e-commerce marketplace.

These companies have a sizable customer base, a large collection, and diverse set of ML use cases, which translates to a large number of ML models. They needed a way to operationalize those ML models efficiently and at scale, and have invested heavily in building their own ML infrastructure, even before the term "MLOps" was coined. As a part of this effort, they stumbled into the MLOps discipline not by choice, but out of necessity.

The earliest glimpse of what MLOps looks like was described in the seminal 2015 pager from Google, "Hidden Technical Debt in Machine Learning Systems,"[19] where it highlighted the ML code is only a tiny portion of real-world ML systems and the surrounding infrastructure is vast and complex, as depicted in Figure 2-7.

---

[18] Thoughtworks, "MLOps Platforms Comparison Matrix" google sheet, `https://docs.google.com/spreadsheets/d/1nRqjnD7SCMJGmYR2gdZJ84YolLnHAMJwjSG7z7VcM6c`

[19] D. Sculley, G. Holt, D. Golovin, E. Davydov, T. Phillips, D. Ebner, V. Chaudhary, M. Young, J. Crespo, and D. Dennison, "Hidden Technical Debt in Machine Learning Systems", NIPS, (2015) `https://proceedings.neurips.cc/paper/2015/file/86df7dcfd896fcaf2674f757a2463eba-Paper.pdf`

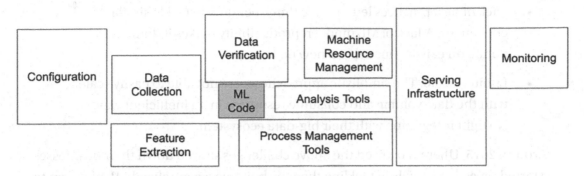

Figure 2-7. *Vast and complex infrastructure around ML code. Source: Adapted from an image in Hidden Technical Debt in Machine Learning Systems paper*

The following sections will showcase two successful and widely known ML platforms: Uber Michaelangelo and Meta FBLearner. These two platforms are designed to support ML operationalization at scale and supported by a large team. The goal here is not to replicate them in their entirety, but rather to borrow their best practices and be aware of the learned lessons.

Uber Michelangelo

One of the most publicized and well-known ML platforms is Uber Michelangelo. Its journey and progression can be easily found on the Internet. Between the Uber ride sharing marketplace business and the Uber Eats food delivery business, there are many important ML use cases that play a critical role to their business, such as dispatch, dynamic pricing, demand forecasting, ETA, restaurant preparation time, fraud, and much more. The number of ML models in production at any point in time is in the thousands.

During the explosive growth period of Uber's history, around 2015, there was not a centralized way to develop and deploy ML models, which led to

- Fragmentation: Multiple teams were building their own solutions and as a result numerous duplication of efforts.

- No leverage and standardization: Silo ML systems built in their own way were not conducive to enabling leverage across teams.

- Lack of best practices led to more time spending to tackle similar challenges. A lack of ML model reproducibility makes it difficult to iterate on existing models on new data.

- Limited scale: The simplistic approaches were not able to easily scale with the data volume and consume resources in an inefficient way. No tight integration with their big-data ecosystem.

Around 2015, Uber recognized the above challenges slowed down their businesses and started investing heavily in tackling them by building a centralized ML platform to help with productionalize ML. This large undertaking includes establishing a centralized team that is staffed with a product manager and an engineering organization to build Uber Michaelangelo as a first class internal product at Uber.

According to Achal Shah, an ex-engineer on the Michelangelo team, the approaches[20] they took to address the aforementioned challenges and to democratize ML at Uber are the following:

- Build and provide standardized workflows and tools with good out-of-the-box experience and flexibility.

- Provide a set of standardized and performant ML algorithm implementations.

- Provide scalable end-to-end ML workflow to meet various large ML use cases.

- Democratize and accelerate ML through ease of use.

Over the course of a few years, the Michelangelo team was able to achieve most of their goals with a comprehensive and centralized platform to address the end-to-end ML workflow, from data preparation to both online and offline predictions and monitoring, as depicted in Figure 2-8. At the high level, the Michelangelo platform is built with a mix of open source and in-house solutions and is tightly integrated with Uber's data compute infrastructure.

---

[20] Achal Shah, "Michelangelo: Uber's machine learning platform" (2018), www.youtube.com/watch?v=hGy1cM7_koM

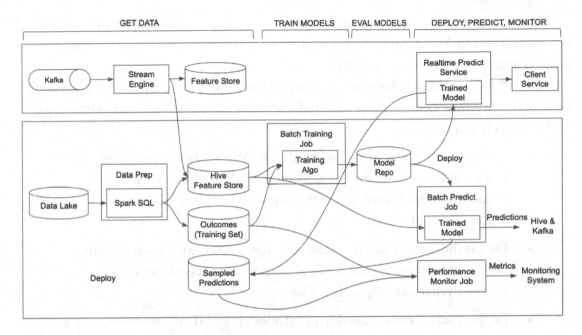

Figure 2-8. *Michelangelo System Architecture. Source: Adapted from Meet Michelangelo: Uber's Machine Learning Platform blog at* `www.uber.com/blog/` `michelangelo-machine-learning-platform/`

One of the key principles that the Michelangelo team adopted is to treat ML development as similar to software development. This implies treating ML artifacts as code, such as feature pipelines, model training pipelines, model development and deployment metadata, and more. These artifacts follow software engineering best practices that include versioning, code review, and rigorous testing. A specific example is about evaluating their ML models against holdout sets before deploying them to production to ensure their performance is not that different from the one in the offline evaluation.

Key Takeaways and Lessons Learned

Building an ML Platform from the group up to support a large and diverse set of use cases for a large ML practitioner community at Uber is a large investment and requires a dedicated team to build, iterate, and support it. The benefits are also very meaningful and impactful because ML plays a critical role in their online marketplaces to drive their business forward, improve customer experience, and reduce cost.

Without a doubt, one of the key takeaways is that tooling and infrastructure play a key contributor to drive high efficiency and increase velocity in the ML development process and lower the barrier to entry and experimentation.

In the "Michelangelo - Machine Learning @Uber"[21] presentation, Jeremy Hermann, head of the Machine Learning team, shared a few lessons learned from their journey of building Michelangelo:

- To increase ML development velocity, let ML practitioners use the tools that they want to use or most comfortable with, as well as reduce friction at every step of the complex and iterative workflow via automation or tooling.

- Data is the oxygen for ML. Make sure to provide the necessary tooling and infrastructure to enable ML practitioners to access, compute, and analyze data easily and quickly.

- Enable end-to-end ownership of ML models by ML practitioners.

- Provide visual tools for understanding data and models, and make the deployment process fast and easy by hiding the details in the UI.

- ML platform, like Michelangelo, is a large and complex project. Having the long-term vision is one thing, but developing it iteratively based on user feedback will dramatically increase the chance of success.

Meta FBLearner

Meta (formerly known at Facebook) is one of the pioneers in applying ML to provide unique and personalized experiences to their large user base on the Facebook, Instagram, Messenger, and WhatsApp platforms. Their diverse set of use cases include personalized news feed stories, filtering out offensive content, ranking search results, language translation of more than 2 billion stories everyday so their users can connect in any language,[22] advertisement recommendations and serving, speech recognition, content understanding, and many more.

---

[21] Jeremy Hermann, "Michelangelo - Machine Learning @ User" (2018) – www.infoq.com/presentations/uber-ml-michelangelo/

[22] "Machine Learning at Meta," 2022 – www.metacareers.com/life/machine-learning-at-facebook/

Meta's approach to building their own ML infrastructure is quite unique from most other companies due to the large scale they are operating at, as well as having a very large and diverse set of use cases. However, their journey of getting to a mature ML infrastructure doesn't deviate too much from the journey of other companies.

Effective machine learning has a large dependency on data, and Meta collects a vast amount of data from their vast user base who interact on their platform on a regular basis. On a daily basis, multiple petabytes[23] are generated from billions of stories and hundreds of million uploaded images. From the infrastructure's perspective, they recognize a need to build a massive infrastructure to efficiently store, move, and compute the large volume of data so their researchers, data scientists, and engineers can easily and quickly access the data to train ML models on a daily basis. Due to the large data volume and scale that Meta operates at, the challenges, as depicted in Figure 2-9, are daunting and require large investment, manpowers, and unique and innovative approaches.

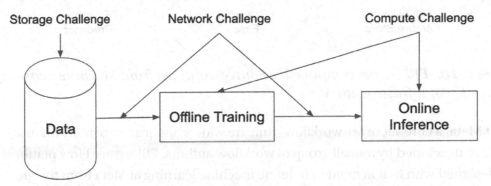

Figure 2-9. *Infrastructure challenges with building and scaling AI/ML systems. Source: Adapted from Machine Learning at Facebook: An Infrastructure View[24]*

In addition to designing and operating their own data centers around the world, they also design specialized hardware to support the various ML-related workloads, including model training, model evaluation, online inference, and more.

[23] Janet Wiener, Nathan Bronson, "Facebook's Top Open Data Problems" (October 2014), https://research.facebook.com/blog/2014/10/facebook-s-top-open-data-problems/

[24] Yangqing Jia, Machine Learning at Facebook: An Infrastructure View, 2018, www.youtube.com/watch?v=qTRLabsOOhQ

At Meta, one of their primary goals of building ML infrastructure is to democratize ML such that engineers without a strong ML background can iterate quickly on existing ML models to improve their accuracy and quickly and easily deploy those models to production with guardrails. The main driving force behind this is their large product surface can benefit from ML and their engineer population is way larger than the data scientist population; in other words, there aren't sufficient data scientists to meet the need to apply ML.

In late 2014, their ML infrastructure journey started with the construction of FBLearner to provide a common support for building, training, deploying, and serving ML models across the company, as depicted in Figure 2-10. The Feature Store component provides an easy way to store and access features, the flow component provides a facility to develop and train ML models, and the Predictor component provides online predictions at scale.

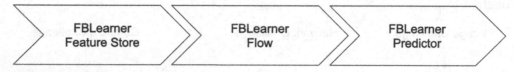

Figure 2-10. *FBLearner components. Source: Adapted from Machine Learning at Facebook: An Infrastructure View* [26]

At Meta, a core set of ML workflows that are widely applicable to many ML use cases are developed by a small group of workflow authors. FBLearner Flow platform was designed with that in mind to redefine machine learning at Meta from the ground up, and to put state-of-the-art algorithms in AI and ML at the fingertips of every Meta engineer.[25] The aim is to make it easy to reuse and extend those workflows, to run thousands of simultaneous custom experiments and manage ML experiments with ease. The FBLearner Flow platform, as the workflow management system, consists of three core components: an authorship and execution environment for distributed workflow, an experiment management UI for launching and analyzing experiments, and a set of predefined ML pipelines that serve as a starting point for training the most commonly used ML algorithms at Meta.

---

[25] Jeffrey Dunn, "Introduction FBLearner Flow: Facebook's AI backbone" (May 2016), https://engineering.fb.com/2016/05/09/core-data/introducing-fblearner-flow-facebook-s-ai-backbone/

To achieve those goals, the FBLearner Flow team adopted the following guiding principles: reusability, ease of use, and scale.

The reusability principle is accomplished by providing the building blocks that can be composed in a reusable manner, which are workflows, operators, and channels. Each instance of a workflow is called a pipeline, and it is created for a specific ML use case, and is made up of a set of ML-specific tasks, such as training and evaluating a specific ML model. Each task is defined in terms of operators that can be chained together and can be executed in parallel. Operators are the smallest unit of execution. Each pipeline typically needs some inputs to feed into the model training and the pipeline outputs are persisted somewhere, and the inputs and outputs are represented as channels.

The productivity principle is accomplished by the powerful and flexible UI of the experiment management component for customizing, launching, and visualizing the ML pipelines. This component was designed to interpret workflow definition and inputs and outputs in a generic manner, as well as to provide a plugin system for further customization to meet the needs of the various teams and easily integrate with internal Meta's systems.

The comprehensiveness principle is accomplished by providing a consistent and rigorous workflow validation and a toolset that supports diverse needs around the various ML libraries.

As a result, FBLearner Flow gained a wide adoption where over 25% of Meta engineers have used it to deliver personalized experiences to Meta's large user base and have trained more than one million models.

Key Takeaways and Lessons Learned

The journey Meta started in 2014 to build out the FBLearner platform has proven to be an effective and scalable way to democratize ML. However, that was just the beginning, and it will require additional and continuous effort and investment to build out the remaining pieces to address the entire ML development cycle and to fully adopt the MLOps best practices in order to increase the velocity of applying ML.

The following learnings were shared by Aditya Kalro, a Senior Software Engineer Manager at Meta, in the "Evolution of Machine Learning at Facebook"[26] webcast, conducted in 2020 by Sam Charrington at TWIML.

---

[26] Sam Charrington, Aditya Kalro, "The Evolution of Machine Learning Platforms at Facebook" (2022), www.youtube.com/watch?v=IOE43Up2L7k

- Treating ML development using lessons learned from software development. ML Model development and productionalization have similar needs as software development, such as a smooth and efficient build and release process to make it easy to retrain models, model monitoring and debugging are important once the models have significant contribution to the user experience across many products.

- ML lifecycle consists of many stages, and they all need infrastructure and tooling to increase the velocity and efficiency of the overall ML development lifecycle. In the beginning, the feature store component of the FBLearner platform provided only the support for storing and accessing features, and the feature engineering development was left to the engineers. For some ML use cases that deal with unstructured data, such as image classification and speech recognition, they can greatly benefit from the labeled data to enable the ML algorithms to learn patterns and make accurate predictions. Infrastructure or tools that help speeding the data labeling process will speed the ML development lifecycle.

Summary

Successful MLOps adoption requires strategies and thoughtful approaches. This chapter starts with a reminder that the MLOps adoption strategies need to be customized to the uniqueness and the various needs of each organization. It then highlights two very critical aspects that organizations need to figure out:

- Alignment between business goals and MLOps infrastructure goals

- Assessment of the specific MLOps needs

One organization's MLOps needs will likely be very different from the next organization. One way to assess the needs is to gather the details about these dimensions: ML use cases, the current state and maturity of the technology that MLOps depends on, the current size and talent in their organization that are related to Data Science and MLOps infrastructure, and most importantly their culture.

Once the MLOps needs are identified, the next step is to determine the most suitable approach to make MLOps a reality. The three common approaches are buy, build, and hybrid. Given a plethora of commercial vendor solutions and open source innovations in the MLOps space, most organizations would likely lean toward either a buy or a hybrid approach. However, the MLOps landscape is growing and evolving rather quickly, organizations need to carefully evaluate the maturity of the commercial vendor solutions and to clearly understand where a particular solution falls in the end-to-end platforms and specialist tools spectrum.

This chapter ends with a showcase of two successful and widely known ML platforms: Uber Michaelangelo and Meta FBLearner. These two platforms are designed to support ML operationalization at scale and were built and supported by a large team. Their best practices and some of the learned lessons are valuable insights for organizations to consider as they venture into their MLOps adoption journey.

CHAPTER 3

Feature Engineering Infrastructure

This chapter and the following infrastructure-related chapters are designed to dive into details about the infrastructure of the MLOps stack, namely, feature engineering infrastructure, model training infrastructure, and model serving infrastructure. For these chapters, the following structure will be used. First, the high-level details and benefits will be described. Next, the high-level architecture and its subcomponents will be discussed, and finally a few case studies, including home-grown, open source, and commercial vendor solutions, will be highlighted.

The hope is the technical details and discussions in these infrastructure-related chapters are useful for MLOps leaders, technical leads, and MLOps engineers to gain a high-level and technical understanding of the intricacies and trade-offs and more importantly to guide in their technical decisions while building out their MLOps infrastructure or evaluating solutions from the open source community or commercial vendors. In addition, a few potential organization-related challenges will be called out when relevant.

Note The term "infrastructure" used in this book is quite similar to the term "platform" being used in other MLOps books or blogs. These two terms are often used interchangeably. For me the term "infrastructure" is more related to the technical aspects, and a platform is a more encompassing and larger area that consists of multiple pieces of infrastructure. That's the reason the MLOps platform (ML platform) is used in this book to cover the entire spectrum of needed infrastructure to support the ML development process.

© Hien Luu, Max Pumperla and Zhe Zhang 2024
H. Luu et al., *MLOps with Ray*, https://doi.org/10.1007/979-8-8688-0376-5_3

Overview

One of the first and critical steps in the ML development process is the feature engineering step, where data scientists select and transform raw data into a set of features that will be used to train and build a machine learning model. The quality and the right set of features can greatly influence the accuracy and performance of the ML model. Feature creation and generation, especially at scale, have a large dependency on the data infrastructure maturity and have numerous engineering challenges. These are some of the reasons why data scientists spend a significant amount of time in the feature development step when there is no feature engineering infrastructure to support them. Like most other steps in ML development, feature engineering is an iterative process to refine or develop new features once new insights are learned from experimentation.

Before diving into the details and specific capabilities of feature engineering infrastructure, let's first understand the feature engineering process and the needs from the data scientist's perspective.

The feature engineering process might vary a bit depending on the specific problem that needs to be solved with machine learning, but the general steps data scientists follow through are

- Feature discovery: A quick and easy way to speed up the feature engineering process is to reuse existing features if they are available and suitable for the use case at hand. As an organization scales the number of ML projects, undoubtedly a set of base features will emerge, and oftentimes they are useful for similar ML use cases.

- Exploratory data analysis: Identify and explore the data to pick out potential features. This involves analyzing and visualizing data, identifying patterns, and understanding any data quality issues.

- Feature transformation: From the identified data, generating features by transforming it using mathematical operations, statistical measures (i.e., mean, mode, variance), or various feature engineering techniques, such as one-hot encoding, scaling and normalization, imputation, dimensionality reduction, or text processing.

- Feature selection and validation: From the various generated features, identify the most relevant and useful ones to train the ML model and then check for overfitting and underfitting.

- Feature serving: This step is needed during model training and online inference.

- For online ML use cases, the features will be refreshed on a regular basis or at real-time and will need to be available at inference time in low latency.

Note   Data vs. feature

A common and reasonable question often asked is what features are and how they are different from data. In the context of machine learning, features are measurable properties or characteristics of an object or entity that can be used as input signals for models to learn patterns from. Features are created from data via a transformation process called feature engineering. Data is a more general term that refers to all of the information.

At the high level, feature engineering infrastructure enables data scientists to quickly and easily develop the relevant and high impact features without dealing with the underlying engineering complexities so they can bring their models to production in the shortest amount of time.

Each infrastructure piece in the MLOps stack plays an important role in their own way to help bring ML models from idea to production quickly. One important and unique aspect that elevates the importance of the feature engineering infrastructure above the others is its output, namely, features, is the main ingredient for ML projects and is used in multiple phases of the ML development process. They are model training and model inference, and therefore getting this component right is quite important as organization scales up their ML projects.

To enable data scientists to mainly focus on the feature engineering process, this infrastructure provides all the necessary abstraction, toolings, automation, storage, and more, such that most if not all the engineering aspects are abstracted away, hidden, and taken care of. The separation of concerns about what features to generate from the various data sources and how to generate them in a consistent and efficient manner, at scale, and at certain cadence is the central purpose of the feature engineering infrastructure.

Benefits

The benefits the feature engineering infrastructure brings will be more obvious and magnified when an organization scales up the number of production ML use cases, when there are multiple teams applying ML to business problems, or when the feature reuse topic comes up frequently during ML reviews or discussions.

Among the other benefits, these are the common ones that a feature engineering infrastructure provides:

- Model performance

  - For online ML use cases, the infamous training-serving skew problem can cause the model's performance to degrade significantly, leading to poor predictions, decreased accuracy, and ultimately impact customer experience or reduce business impact. The feature consistency between training and inference is provided by this infrastructure.

- Unlock online ML use cases

  - Online ML use cases, such as personalized recommendations, fraud detection, etc., require features to be available at inference time with low latency retrieval time and near real-time freshness. The feature engineering infrastructure enables the ability to serve features at low latency reliably and efficiently and therefore enables organizations to leverage ML in more and impactful use cases.

- Productivity and efficiency

  - Feature engineering infrastructure provides necessary tooling and infrastructure to automate the feature engineering process, which reduces the time and effort required to create, manage, and maintain features. This frees up data scientists to focus on other tasks such as model selection, tuning, and deployment.

- Collaboration

  - By treating feature definition as code, data scientists and engineers can easily collaborate on feature creation and generation, share best practices, and reuse feature sets across different models and projects.

- Governance

  - With a centralized way of managing features, feature engineering infrastructure provides the necessary infrastructure to enforce governance policies, such as data privacy and security, data lineage, and version control for an organization.

Note   Training-serving skew

This infamous problem typically shows during the model production step, and it causes model performance degradation in an unexpected way. A good and clear definition of this problem can be found in this Rules of Machine Learning resource.[1] The skew part refers to the model performance difference between the training phase and production phase, and one common cause to this difference is about the way the features were computed or generated.   Typically, the model performance is expected to be at the same level or better once a model is in production.

High-Level Architecture

A good feature engineering infrastructure should be able to meet most if not all the needs of the feature engineering process that data scientists regularly go through. The specific capabilities to satisfy those needs include the ability to discover the existing features, create new ones when necessary, manage feature access and lifecycle, orchestrate and transform raw data into features, monitor the feature freshness and quality, and ensure those features are reliably and consistently available at inference time at low latency.

---

[1] "Rules of Machine Learning: Best Practices for ML Engineering," https://developers.google.com/machine-learning/guides/rules-of-ml

From the engineering's perspective, this large surface infrastructure consists of the following sub-components:

- Feature catalog acts as a feature metadata repository to help with feature discovery and various feature management capabilities.

- Feature engineering framework provides an easy way to perform feature transformation and computation.

- Feature store provides a central place to persist the features for model training and model inference purpose as well as a way to easily access features.

- Feature insights and quality provides an up-to-date information about the feature usage and their quality.

- Feature upload service to upload only the needed features to online feature store.

Figure 3-1 depicts at the high level what a feature engineering infrastructure looks like. The upcoming sections will dive into more details about the specific capabilities listed inside the "Feature Engineering Infrastructure" box. For now, let's walk through a few important topics that are not obvious from the figure.

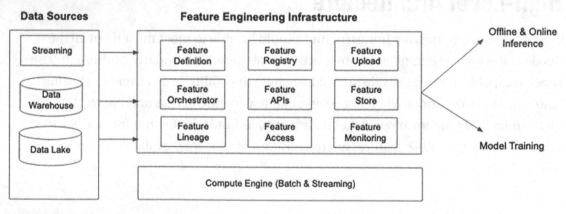

Figure 3-1. *Feature engineering infrastructure high-level architecture*

The most common data sources for creating features are usually the data stored and managed in the company's centralized data warehouse and/or data lake. This is where it is quite important to have a fairly solid data infrastructure to maintain and keep the data up to date in these data sources with high quality. Without this or the access, data scientists will have a very difficult time in applying ML to add value to organizations, unless those ML projects are at the exploratory stage and require only a small and static dataset.

The data from the streaming data sources enables data scientists to leverage real-time signals about customer behaviors or fresh data to accurately predict and quickly adjust to changing conditions, such as in financial markets or real-time monitoring systems. The engineering complexity involved in supporting real-time features is significantly higher, multiple times that of the other two data sources. It would be wise to understand the added values from these features before investing in the time and investment in supporting them.

A good data infrastructure typically provides distributed and scalable compute engines that are easy to use for batching and streaming computation. Feature computation and generation can greatly benefit from the availability of these engines. This is another indication that a modern feature engineering infrastructure heavily depends on a good data infrastructure and is often described as the ML-specific data infrastructure.

The small boxes under the "Feature Engineering Infrastructure" title represent an ideal infrastructure to support a large set of ML use cases across multiple teams. Not all of them need to be checked or fulfilled in order for data scientists to start creating and generating features when there are just a handful of ML use cases.

Note   Feature store vs. feature engineering infrastructure or feature platform

In the context of feature engineering infrastructure, aka feature platform, a feature store is one of the sub-components, and it acts as a central place to store and serve features that have been computed. An example of an open source feature store is Feast (`https://feast.dev/`).

Feature engineering infrastructure represents a broader and more comprehensive solution that consists of a feature store and many other sub-components to manage the complete lifecycle of ML features, from generation to production.

The following sections will briefly describe capabilities of the feature engineering infrastructure subcomponents and highlight some of the common approaches from the industry and open source communities.

Feature Specification and Definition

In the initial phase of an ML project, data scientists typically employ quick and unstructured ways to create features for their ML models after analyzing and visualizing data that potentially will be used as features using tools like Pandas or Spark. As a part of productionalizing their ML models, those features that were used to train those models will need to be generated and materialized in a repeatable and consistent way. This is where the feature specification and definition come in.

They provide a standard and consistent way for data scientists or ML engineers to define features, to express the necessary transformations, and to specify the orchestration details. The feature specification and definition can broadly be grouped into a higher concept called metadata. The following sections discuss the details about the feature metadata and various commonly adopted formats for expressing it.

Feature Metadata

The feature metadata is meant as self-contained metadata to express all the necessary information as a part of the feature engineering step for both human and machine to understand and reason about. It typically consists of the following logical parts: feature metadata, transformation metadata, and specification metadata. The first one refers to the general information about data source to use to generate the defined features and materialization destination to store the computed feature value, and more. The second one refers to the logic to transform the raw data from one value to another, such as converting timestamp to day of week, calculating a mean of the temperature column, etc. The last one is typically reserved for expressing the information about the environment as well as the orchestration.

The list below contains the common metadata that are generally needed:

- Source: Information about the data sources to generate features from. There can be more than one source when there is a join involved. Possible sources are data warehouse, data lake, or streaming source like Kafka.

- Basic information: Entity, name, type, transformation, dates.

- An entity is a logical concept to represent a business domain, such as customer and merchant. It is useful as a grouping mechanism to organize a set of related features under the same entity.

- Each feature is identified by a unique name.

- Each feature has a data type.

- Transformation logic. From simple or standard mathematical transformations to complex aggregation.

- Sink: Once the feature computation is done, where to materialize the feature to. This can be a table in the data warehouse, a certain path in the data lake, or a topic in Kafka.

- Description: Contains high-level context information about the feature.

- Author information: Typically include name and email. Useful for sending failure notifications to or for others to know whom to contact to learn more about the feature.

- Possible information in specification.

  - The compute engine to use to run the transformation logic.

  - The frequency of feature generation, that is, hourly, daily, etc.

  - Online feature store freshness level and upload frequency.

The above list of information is not meant to be comprehensive. They are commonly supported in the various in-house, open source, and vendor feature engineering solutions. Due to the specific use cases, evolving technologies, and human preferences, the naming and semantic might vary slightly between those solutions.

The feature metadata from the in-house solutions, such as Palette[2] from Uber, Feathr[3] from LinkedIn, Chronon from AirBnb, and Fabricator[4] from DoorDash, and open source solutions such as Feast[5] have a lot in common, but they all use slightly different terminologies and concepts.

From the perspective of the feature engineering step in the ML development lifecycle, the needs are quite similar, and it would be a great win for the community if there was a single specification that everyone would stand behind and benefit from.

Feature Metadata Format

There are multiple options at the disposal for the feature engineer infrastructure team to explore and consider. As usual, it is important to keep the primary user persona in mind when making decisions among the various options. Over time, the popular formats emerged, and the not so good ones were submerged after their insufficient sides were visible. It would be easier if there is a single format that rules them all and we all can simply adopt it. The conventions are using YAML format and custom Python API, and each one has its own advantages and disadvantages.

YAML (YAML Ain't Markup Language) is a text-based human readable data serialization language and commonly used for configuration files and can succinctly described as a versioned language for data.[6] It has become a popular format due to its ease of use, versatility, portability, and flexibility. An example of ease of use is that a YAML file can be written using your favorite text editor, such as sublime, visual code, or vi. The main thing that data scientists need to be familiar with is the defined schema for the feature metadata. The schema, which defines the syntax and structure of the file, is typically defined by the creator of the feature engineering infrastructure to specify the structure and the allowed fields in the feature metadata YAML file.

---

[2] "Accelerating ML at Uber with Michelangelo Palette," Amit Nent, 2022 `https://content.hopsworks.ai/hubfs/Feature%20Store%20Summit%202022/FS%20Summit%2022%20-%20Uber.pdf`

[3] "A scalable, unified data and AI engineering platform for enterprise," `https://feathr-ai.github.io/feathr/`

[4] "Introducing Fabricator: A Declarative Feature Engineering Framework," Kunal Shah, 2022, `https://doordash.engineering/2022/01/11/introducing-fabricator-a-declarative-feature-engineering-framework/`

[5] Feast – `https://docs.feast.dev/`

[6] YAML Specification, `https://github.com/yaml/yaml-spec`

Python is one of data scientist's favorite programming languages, and it is no surprise that more and more recent feature engineering solutions adopt this approach for expressing the feature metadata.

While it is extremely easy to start writing YAML files, it is not as straightforward and fast to get feedback when there are mistakes, compared to a Python editor. The Python editor's code completion feature can help with reducing mistakes or quickly lookup the forgotten or unfamiliar options. On the other hand, running and testing a Python file require the availability of a Python environment. Given Python programming language is something that data scientists use daily, there is no longer a large inconvenience.

Regardless of which feature metadata format is adopted, the main goal here is to treat feature engineering artifacts, typically written by data scientists or ML practitioners, as code. This will facilitate the adoption of software engineering best practices such as code review, reproducibility, CI/CD integration, and other automations. This simple, yet so beneficial, concept is one of the steps to increase the ML development velocity.

Another key important concept to highlight is that feature metadata provides a clean and concrete separation of concerns between a data scientist's responsibility and the underlying infrastructure's responsibility to take care of engineering complexity and hide the underlying implementation so they can be evolved easily as more optimal approaches or new solutions are discovered.

YAML-Based Feature Specification and Definition Examples

Some of the in-house solutions are using YAML as the feature specification and definition. Among them are Snap and DoorDash. Listing 3-1 is an example from Snap's Robusta feature engineering framework,[7] which was built to streamline and accelerate ML iteration by reducing frictions around extracting signals for their recommendation systems.

Listing 3-1. Snap's Robusta feature aggregation adaptation

```
name: my_feature_spec
query:
  sql: > SELECT snap_view_spec > 1 as view_time_gt_1,
          snap_id, hour_of_day(event_timestamp),
          day_of_week(event_timestamp) = 'SUNDAY' as is_sunday
```

---

[7] "Speed Up Feature Engineering for Recommendation Systems," September 2022, https://eng.snap.com/speed-up-feature-engineering

```
          ...
        FROM
          discovery_snap_view_data
        WHERE
          event_name = 'DISCOVERY_SNAP_VIEW'
features:
  base_name = discover_snap_total_viewed_Counts
  aggregation:
    count: {condition_columns: [view_time_gt_1]}
  group_by_selectors:
    snap_id: DOCUMENT_ID
    hour_of_day: HOUR_OF_DAY
  primary_select: DOCUMENT_ID
  window_to_ganularity:
    six_hours: five_minutes
    thirty_days: twelve_hours
```

A few interesting design decisions to accommodate the flexibility to call out are:

- The query section accepts a SQL statement as the transformation logic. Given that most data scientists, ML engineers, and data engineers are already familiar with SQL, this should be easy for them to adopt with minimum friction.

- Separate feature aggregation sections allow for different aggregation keys to apply to the same single SQL statement defined above the "query" field.

The other YAML-based feature metadata format example comes from DoorDash's declarative real-time feature engineering framework called Riviera.[8] This example is about computing real-time features from a streaming source to compute a store-level feature that provides the number of orders by store using a sliding window with a length of 30 minutes and a sliding interview of 1 minute.

---

[8] "Building Riviera: A Declarative Real-Time Feature Engineering Framework," 2021, https://doordash.engineering/2021/03/04/building-a-declarative-real-time-feature-engineering-framework/

Listing 3-2. DoorDash's real-time feature aggregation example

```
source:
  - type: kafka
      kafka:
      cluster: ${ENVIRONMENT}
      topic: store_events
      schema:
      proto-class: "com.doordash.timeline_events.StoreEvent"

sinks:
  - name: feature-store-${ENVIRONMENT}
      redis-ttl: 1800

compute:
  sql: >-
    SELECT
    store_id as st,
    COUNT(*) as saf_sp_p30mi_order_count_avg
    FROM store_events
    WHERE has_order_confirmation_data
    GROUP BY
    HOP(_time, INTERVAL '1' MINUTES, INTERVAL '30' MINUTES),
    store_id
```

Similar to the other example, the feature transformation logic is expressed in a SQL statement. SQL has become the lingua franca for working with data, but with evolution, there are multiple SQL dialects, and certain compute engines can support only certain dialects. If the solution being built is an in-house solution, this is less of a concern; however, if it is a commercial solution, then the API-based option is more future proof to support numerous customer use cases.

Python-Based Feature Specification and Definition Examples

Python is the lingua franca of data science, and therefore most if not all data scientists are comfortable working with Python. A few recent open source solutions and commercial vendor solutions have adopted Python as the approach to defining the feature specification and definition.

One of the recent feature engineering open source solutions in 2022 is called Feathr from LinkedIn. It was developed by LinkedIn and battle tested internally for a few years to power numerous ML use cases, such as Search, Feed, and Ads. Feathr acts as an abstraction layer for defining features as well as a common platform for computing, serving, and accessing features.[9] Listing 3-3 contains a short example about defining features using Feathr python APIs.

Listing 3-3. A simple example of using Feathr Python APIs to define a set of features

```
batch_source = HdfsSource(name="nycTaxiSource",
                path="<path>",
                event_timestamp_column="lpep_dropoff_dt",
                timestamp_format="yyyy-MM-dd HH:mm:ss")
f_trip_distance = Feature(
    name="f_trip_distance", feature_type=FLOAT,
    transform="trip_distance",
)
f_trip_time_duration = Feature(
    name="f_trip_time_duration", feature_type=FLOAT,
    transform="f_trip_time_duration",
)
features = [
    f_trip_distance,f_trip_time_duration,
    Feature(name="f_day_of_week", feature_type=INT32,
            transform="dayofweek(lpep_dropoff_datetime)"),
]
request_anchor = FeatureAnchor(name="request_features",
                              source=INPUT_CONTEXT,
                              features=features)
```

---

[9] David Stein, "Open sourcing Feathr - LinkedIn's feature store for productive machine learning," 2022, https://engineering.linkedin.com/blog/2022/open-sourcing-feathr---linkedin-s-feature-store-for-productive-m

The above example shows the Feature class is used to define the information about a feature, which includes name, type, and optional transformation logic. More details can be found at this resource – https://feathr-ai.github.io/feathr/concepts/feature-definition.html.

One of the advantages of using Python APIs is to be able to define features in small and composable units and when needed, compose them together to form a complex feature pipeline.

Another fairly popular and widely adopted open source feature store project is called Feast. Its feature definition approach is quite similar to the one from Feathr, using a Python class abstraction for definition features; see Listing 3-4 for an example.

Listing 3-4. A simple example of using Feast Python APIs to define a set of features

```
driver = Entity(name="driver", join_keys=["driver_id"])

driver_stats_fv = FeatureView(
      name="driver_activity",
      entities=[driver],
      schema=[
         Field(name="trips_today", dtype=Int64),
         Field(name="rating", dtype=Float32),
      ],
      source=BigQuerySource(
         table="feast-oss.demo_data.driver_activity"
      )
)
```

In Feast, each feature is defined using a class called Field, and a collection of related features is logically grouped together in an instance of the FeatureView class.

The above examples are meant to illustrate the two common formats used for feature definitions.

The feature format is a means to an end. The ultimate goal of the feature specification and definition is to make it easy and quick for data scientists to perform feature engineering in a consistent and standard way. Independent of the chosen format, we can easily treat the feature specification and definition as code, and therefore, we can easily benefit from some of the DevOps best practices, such as versioning, code reviewing and reproducibility, and more.

Another important benefit that the feature specification and definition bring is the consistency during offline and online feature serving because the feature transformation logic is defined once and in a centralized place.

Once the feature specification and definitions are defined and available, what happens next?

Feature Registry

Once the feature specifications and definitions are code reviewed and checked into a versioning system like GitHub, the details and metadata about the defined features are compiled and converted into an internal format, and persisted into a feature registry.

The feature registry acts as a central catalog of all the features for an organization or team. One very important benefit it brings is to enable data scientists and other ML practitioners to search, discover, and collaborate on new features. In addition, it can provide additional useful capabilities such as feature level lineage data and access control.

Indirectly, it promotes feature reuse through discoverability. When data scientists first join a team or embark on a new ML project or improve an existing ML model, one of the first actions they take is to get a sense of the available features that are being used in production.

From the technical perspective, the feature registry is typically designed as a web-based application that is made up of the following components: backend, web app, and user interface (Figure 3-2). The backend acts as a repository for storing the feature specifications and definitions and is commonly designed as SQL based. The web app is for managing the CRUD operations of those artifacts, and the user interface exposes those same artifacts and facilitates the management and control.

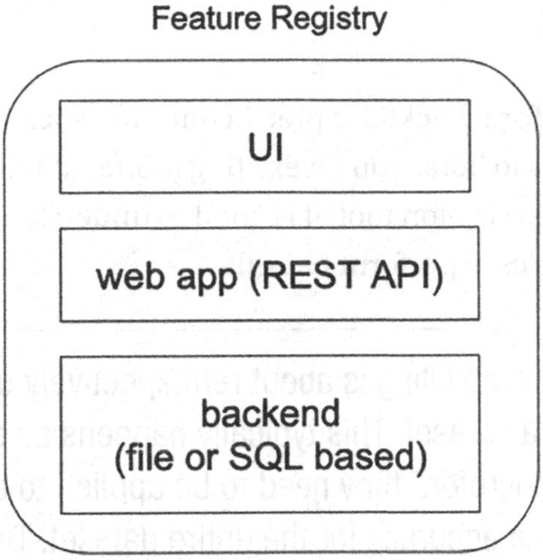

Feature Registry

Figure 3-2. *Feature registry components*

Feature Orchestration

The feature specifications and definitions essentially capture the "what" intentions from the ML practitioners in terms of what features need to be generated from what data sources and the specified transformation logics to convert the raw data into features.

Underneath the hood of the feature engineering infrastructure, the "what" intentions are converted into the "how" by translating the feature specifications and definitions into a series feature pipelines and then scheduled and run on a certain cadence. The output of this conversion process is a series of jobs that are formed in DAG structure, which is then managed and executed by the feature orchestrator.

From the technical perspective, feature orchestration is mainly about using a workflow and scheduling system, such as AirFlow, Dagster, etc., to schedule and execute feature pipelines that are represented as DAGs of data pipelines and feature specific jobs.

Once the feature pipelines are completed, the output or result is published to a feature store, which will be discussed next.

Key Considerations

- The need to perform backfill is pretty common, such as adding a few new features to iterate on an existing model. When deciding on a feature orchestration tool, it is good to understand how much support it provides to perform backfill.

Note At the high level, backfilling is about retrospectively updating for filling in historical data within a dataset. This typically happens as enhancements are made to a dataset, and therefore they need to be applied to existing historical data to ensure consistency and accuracy for the entire dataset. The backfill process usually entails reprocessing the past data.

Feature Store

One of the critical components of the feature engineering infrastructure is the feature store, which is responsible for storing and serving features for model training and model inference purpose, as depicted on the right part of Figure 3-1.

The output from the feature generation step is a collection of feature values. During the exploration phase of the ML projects, the total size of feature values might be small and usually materialized or persisted on a data scientist's laptop. During the model training and productionalizing step, the feature values are persisted in a feature store.

During the model training phase, the feature values are computed from months' or years' worth of data, depending on the needs of the use case. The offline feature store is designed for supporting this need.

For online inference use cases, the feature values are needed at model inference time, and this is where the online feature store comes into the picture.

Offline Feature Store

The offline feature store plays a critical role during the model training and evaluating phase by helping data scientists to store and serve a large amount of computed features in an efficient manner by providing a centralized, scalable, and efficient repository of features that can be easily accessed and used across multiple ML models.

The offline feature store typically sits on top of a data warehouse or data lake that is backed by a distributed storage system such as S3, Snowflake, Redshift, or BigQuery. For ML use cases where the feature value volume is large in terabytes, a data lake is preferred due to the ability to access data in a distributed manner using various distributed computing engines.

For large ML use cases, such as recommendations or personalization, that need a large amount of features to train with, such capabilities and flexibility will help with reducing training time from days to hours.

Key Considerations

- Leverage the best practices and tooling of the modern data infrastructure to build an offline feature store, such as using binary data format, good data-partitioning strategy, and retention policy.

- An efficient offline feature store is typically co-located with the central data lake to minimize the data transfer, which translates to cost saving.

Online Feature Store

Nowadays, more and more organizations are incorporating ML into their online products or services to perform online inference to either recommend products to customers, or to provide customized browsing experience, or to detect fraud in near real-time. All these ML use cases have a common need, which is about fetching a subset of the computed feature values at low latency and at high QPS.

Online feature stores are designed to meet these needs by keeping those features in distributed key-value stores, such as Redis, Cassandra, or DynamoDB.

Key Considerations

- In-memory key-value stores like Redis can be expensive while storing a large amount of features. Upfront investment in designing an efficient schema for feature values will save costs and minimize the migration effort.

- Be mindful about updating online feature store to ensure minimum impact to the online ML use cases require low latency feature retrieval during online inference.

- Not all online ML use cases require a single digit latency feature retrieval time. For use cases that can tolerate ~25ms feature retrieval time or more, a disk-based storage engine is a viable solution to help be being cost efficient.

Feature Upload

For a large set of ML use cases that require an online feature store and the feature volume is significant, there will be a need to build or adopt an efficient and smart feature upload tool to upload those feature values to an online feature store.

Key Considerations

- Time efficiency: Given a large volume of feature values, the upload tool must be designed with parallelism in mind.

- Upload strategy: Certain feature sets need to be freshed more frequently than others. The upload tool must be designed to accommodate the diverse freshness needs.

- Operational excellence: For certain online prediction use cases, feature freshness plays a big part in the model performance. A well-designed feature upload tool must produce accurate and detailed metrics in order to detect and monitor the feature value freshness and upload status of each feature set.

Feature Serving

Loose coupling is considered to be a good software engineering practice. The feature serving component is typically designed as a service that exposes endpoints to fetch features from the online feature store and hides the underlying storage solutions, such that when introducing a more efficient and faster solution or supporting multiple storage solutions would not cause major disruptions to online feature consumers.

Some of the benefits from this approach are

- Perform additional feature computation or transformation at feature serving time
- Make it easier to evolve or swap the online feature store with a more cost efficient one or newer technology with minimum disruption to feature value consumers

Monitoring

Once a model is operationalized, the model itself doesn't change until the next iteration. The part that is changing are the features being generated based on the arrival of new data, whether that is daily or weekly.

The needed monitoring in the feature engineering process can be broadly categorized into two areas: feature quality and feature pipeline operation.

Feature quality is mainly about the feature distribution shifting and feature value.

- During the model training phase, the model was trained with the training feature set that has a certain distribution. If the online predictions are computed using the same feature set with a different distribution, then the model performance will be degraded. Therefore, it is critical to monitor the shift in feature distribution as they are being generated.
- The feature values are typically calculated from one or more upstream data sources, which might be owned by one or more teams. When intentional or unintentional data-related changes cause feature quality issues such as missing values, outliers, or inconsistencies, ultimately this will have a direct impact on the model performance.

The feature pipeline operation is mainly about the operational aspects and more related to software engineering. The needed monitoring in this area is to ensure the health of feature pipelines, whether they complete successfully, on time, their efficiency in terms of resource usage, etc.

Overall, monitoring plays an important role ensuring the model continues to perform as expected once they are in production by quickly detecting potential feature quality issues or feature operational aspects and mitigating them early on. The less time data scientists need to spend on debugging model performance degradation due to features, the more time they can spend on developing and experimenting existing or new models.

Build vs. Buy

As mentioned before, feature engineering infrastructure plays a pretty critical role in the overall MLOps infrastructure to help with operationalizing ML models. Getting it right is imperative for organizations as they are looking to scale up with ML projects. A question often comes up is whether to build or to buy it. Similar to other technology adoption decisions, there isn't a single and easy answer. There are many factors that we need to take into consideration to arrive at such an important decision. The following sections will examine some of the important factors as well as discussing the trade-off.

Important Factors

Before putting effort into a build vs. buy decision, a good practice is to clearly survey and identify the needs in the organization and align on the expectations around the benefits that are expected to be gained. Depending on the needs, the benefits might vary slightly, but a common set of benefits from adopting a feature engineering infrastructure are

- ML model accuracy: It is widely understood that feature quality and freshness have direct and positive impact on maintaining the model accuracy once the model is in production. In addition, a consistent way of serving features across offline and online will dramatically reduce the infamous challenge around training and serving skew.

- Team collaboration: MLOps is a team sport, and feature engineering infrastructure can help with facilitating and increasing team collaboration between data scientists, ML engineers, and data engineers with capabilities such as feature reuse, treating feature specification and definition as code, which will make it easy and quick to review and iterate on.

- Unlock online ML use cases using real-time features: Generating and maintaining real-time features through stream processing presents many challenges to organizations without a sound data infrastructure. A good feature engineering infrastructure will help overcome this challenge.

Given each organization has their own specific needs, goals, and perspective, a particular important factor for one organization might be less important to another one. The list of important factors below captures the ones that are fairly common across organizations, and each organization would need to determine their own level of importance when using it as an input signal into the buy vs. buy decisions:

- Time to market: It is no secret that building a feature engineering infrastructure from scratch will take time, effort, and of course engineering resources. If there is a short time window to productionize ML models due to business needs or some business-related reasons, adopting a commercial vendor solution is a much more viable option.

- Cost: Like most other software projects, one of the biggest costs is the manpower. The cost factor might be justifiable if ML is considered an integral part of an organization's competitive advantage or a strategic bet; otherwise, buying pre-built infrastructure is a good starting point.

- Customization and integration: Each organization has their own data infrastructure, CI/CD compute infrastructure, and other existing systems. It is much more seamless when building an in-house feature engineering infrastructure when the level customization is quite high.

- Support: Independent of building or buying a feature engineering infrastructure, maintaining, upgrading, and user support are inherent parts of an infrastructure territory. An organization can just pay for the dedicated support and maintenance when the commercial vendor solution is adopted. Otherwise, the support burden usually falls on the team that is responsible for building the in-house solution.

Build

Most of the Internet companies founded before 2015 have been invested in building their own feature engineering platform, such as LinkedIn, AirBnb, Twitter, Uber, Meta, and more. For them the decision was relatively straightforward due to the nonexistence of both open source and commercial solutions back then.

A few commonalities shared by these companies are that they operate at scale, ML plays an integral part of their online product offerings, their ML use cases are mainly online and greatly benefit from real-time features, and they have a fairly sound data infrastructure.

Given the innovations and availability of open source solutions in the feature engineering infrastructure space, if an organization is deciding to build its own solution, it is beneficial to consider an adopt-then-build strategy.

At this stage in the MLOps adoption curve, it is hard to justify a decision to build feature engineering infrastructure from scratch. A smarter and more acceptable approach would be adopt-then-build option. This means adopting one of the available open source solutions as a starting point, and complementing it with in-house solutions to meet the specifics or to fill gaps.

At the time of writing this, there are a few available mature open source solutions, such as Feathr,[10] Feast,[11] and Hopsworks.[12]

Buy

At the time of writing this, there are at least a handful of vendor solutions available to consider. These solutions are offered by a mix of cloud providers, such as AWS, Google Cloud, and vendors, such as Tecton and Databricks.

Given vendors are very eager to engage, it is a bit easier and faster to find and assess the capabilities of their solutions with minimum effort. A large effort should be reserved assessing which and how vendor solutions best meet the organization's needs in the feature engineering infrastructure area. A few key considerations to keep in mind are

---

[10] "Feathr: A scalable, unified data and AI engineering platform for enterprise," https://github.com/feathr-ai/feathr

[11] "Feast: Feature Store for Machine Learning," https://feast.dev/

[12] "Hopsworks: A data platform for ML with a Python-centric Feature Store and MLOps capabilities," https://docs.hopsworks.ai/latest/

- Vendor lock-in: A common uneasiness arises when adopting a vendor solution. One way to alleviate this concern is to go with vendor solutions that are built on top of open source solutions.

- Piecemeal approach: The flexibility around adopting a vendor solution in a piecemeal fashion to gain confidence while assessing the maturity and ROI. Before commiting to a vendor solution with both feet, it would be good to have the ability to start with dipping in one or two toes first.

- Cost: Clearly understand the cost model, whether that is based on the number of API calls, amount of storage volume, or other cost unit. It is a good practice to establish a cost forecast before signing a contract, unless the initial amount is relatively insignificant.

- Support: Evaluate the different levels of available support and figure out the one that best fits your organization's needs. For example, whether the vendor support hours meet your organization's needs when an emergency comes up.

- Compliance: Top-tier vendors typically can meet most if not all enterprise's security and privacy compliance requirements. It is important to double check, especially when working with solutions from small or young startups.

Organizational Challenges

Up until this point, what has been covered and discussed are around the technical solutions in the feature engineering infrastructure. According to this excellent paper "Socio-Technical Anti-Patterns in Building ML-Enable Software,"[13] there are real organizational challenges that prohibit or reduce the effectiveness and velocity in the feature engineering phase of the ML development lifecycle.

---

[13] Alina Mailach, Norbert Siegmund, "Socio-Technical Anti-Patterns in Building ML-Enable Software," 2023, https://sws.informatik.uni-leipzig.de/wp-content/uploads/2023/01/socio-technical-anti-patterns-icse2023.pdf

Data Availability

At this point, it is evident that data is one of the main ingredients to successfully apply ML to achieve business goals. If the needed data to apply ML is restricted or partially available, then that will hinder the ML effectiveness and ultimately lead to ML project failures.

It is imperative for leaders of the data producers and consumers to sit down together to align on the high-level business goals around the organization's ML initiatives and collectively figure out ways to consistently make data available in a centralized location in a reliable and high-quality manner. A few simple steps to consider:

- Awareness: Data producers and consumers very likely belong to different organizations, and therefore, data producers might not be aware of how their data is being used. Data consumers should take an initiative to raise awareness about the importance of their data.

- Data centralization: As the number of ML projects increases, the greater the need to have centralized data ingestion strategy to reliably move data from various sources inside an organization to a centralized place, such as a data lake or data warehouse.

Data Governance

A commonly held belief among the ML practitioners is that the vast majority of underperforming ML models in production are actually victims of data-related issues, such as low quality, refreshness, missing values, etc.

As there are more and more ML projects productionalized, it is important to protect the down side by prioritizing data governance practices to ensure the data used to train and serve ML model is of high quality and issues are detected early and mitigated quickly.

Data governance is a large topic and consists of many components. The ones particularly relevant to the ML area are:

- Data quality: Establish good practices around automating data quality validation in the data pipelines and only publish datasets once they pass the validations.

- Data lineage: By understanding the data sources that are used to generate features, it will help with speeding up the time it takes the triage and mitigate data-related issues, such as when the data schema was changed or when the value of certain column was changed from text to numeric.

- Data stewardship: Ensure most, if not all, critical datasets have data ownership. This will help mitigating the issues quickly by knowing who to call for help.

Case Studies

The above sections highlight the benefits of having a feature engineering infrastructure, outline its high-level architecture, and discuss the details of the various components. This chapter ends with examining an existing solution from each category: in-house, open source, and commercial vendor. Each one of these solutions will need a whole book in itself to cover their capabilities in detail. The goal here is to assess their capabilities at the high level and along the way point out their areas of strength and gaps when applicable.

Open Source

There are only a few truly open source solutions in the feature engineering infrastructure area. One of the active and fairly popular projects is called Feast (https://docs.feast. dev/). The name Feast is derived from the first few letters of each of these two words: **fea**ture **st**ore.

Overview

In the context of the feature engineering infrastructure, Feast fulfills most of the capabilities in the feature store component. It is designed to be a "customizable operational data system that reuses existing infrastructure to manage and serve machine learning features to realtime models."[14]

---

[14] "Feast Introduction," https://docs.feast.dev/

The three key objectives that Feast aims to achieve are

- Feature consistency across training and serving: Provide tooling and abstraction to manage and access the features in the offline store and online store.

- Avoid data leakage: Ensure point-in-time correct feature values do not leak to models during training.

- Decouple feature store from data infrastructure: Provides a single entry point to abstract the underlying storage engines for storing features that are used for training models and serving models.

Figure 3-3 depicts the role that Feast plays in the landscape of a feature engineering infrastructure, which consists of the three components in the Feast box. The Architect section below will discuss the capabilities of those boxes.

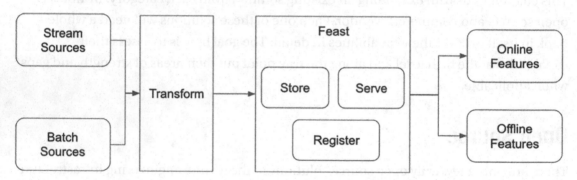

Figure 3-3. *Feast feature store – adapted from* `https://docs.feast.dev/`

Like most other open source projects, Feast is constantly evolving and being improved by the open source community. However, there are certain key aspects that one needs to be aware of when evaluating Feast:

- Feast assumes the feature transformations are already done before you hand them over to Feast to manage.

- Feast provides basic capabilities around feature discovery and has a plugin architecture to integrate with popular data catalog solutions, such as DataHub, Amundsen, etc.

- Feature supports tabular data only and not suitable for unstructured data ML use cases.

Concepts

When working with any infrastructure, tooling, or platform, one of the initial tasks we need to take on is to learn and understand their core concepts, which are the provided abstractions for us to define, use, and interact with.

The concepts in Feast are structured in a hierarchy with three levels. The lowest level consists of Data Source, Field, and Entity, and each one is described here:

- Data Source: Refer to the underlying storage system where the data is resided. Two common types of data sources are batch and stream. Examples of the former one are BigQuery, Snowflake, and RedShift. Examples of the later data source are Kafka and Kinesis.

- Field: A feature, which is often described as an individual measurable property.

- Entity: A container to define a collection of semantically related features. Typically maps to a domain object of particular use case.

The second level has only one concept, and it is referred to as Feature View. This represents a logical group of features originating from a data source and might belong to one or more entities.

The third level contains a concept called Project, which represents the top-level namespace for one or more Feature Views.

Architecture

Feast has been designed as a feature store solution, which is one of the components of the feature engineering infrastructure. Its architecture is pretty easy to grasp, and it consists of a handful of elements, which are depicted in Figure 3-4.

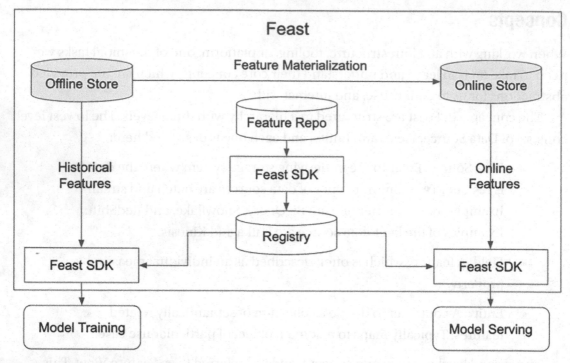

Figure 3-4. *Feast architecture*

One of the key pieces in the Feast architecture is the Feast SDK, which provides an abstraction for the following capabilities:

- Convert the user-provided metadata about the data sources, entities, and features into its internal representation and store them in the registry, which can be backed by various storage engines, such as MySQL or Postgres, etc.

- Perform feature materialization by uploading features into the online store, either from the offline store or pushed into from a streaming source such as Kafka or Kinesis.

- Retrieval of historical features from the offline store for model training purpose with point-in-time correct.

- Retrieval of online features at model serving time.

One area that is not obvious from the diagram above and it is worth calling out is the pluggable design in Feast for operations like materializing data, updating storage engines, launching streaming ingestion jobs, and fetching features from both offline and online stores. This extensibility mechanism will come in very handy when trying to integrate Feast with an organization's internal infrastructure.

Out of the box and through open source contributions, Feast supports some of the most popular and commonly adopted data sources, offline stores, and online stores. See this resource for more details: `https://docs.feast.dev/roadmap`.

Table 3-1. *Supported data sources, offline and online stores*

Data Source	Offline Store	Online Stores
Snowflake, Redshift, BigQuery, Azure Synapse, SQL, Parquet, Hive, Postgres, Spark	Snowflake, Redshift, BigQuery, Azure Synapse + SQL, Postgres, Trino	Snowflake, DynamoDB, Redis, DataStore, Bigtable, SQLite, Azure Cache, Cassandra, Postgres

Assessments

With contributions from the open source community, Feast will continue to evolve into a more complete feature store solution over time. Here is an opinionated summary of the strengths and opportunities based on my observations:

Strengths

- A reasonable foundation and abstraction for feature definition through Python APIs and a registry with pluggable storage engine.

- A built-in support for timestamped tabular data and ability to reproduce the state of features at a specific point in the past via point-in-time joins.

- A reasonable abstraction to upload features to online store from the offline store via the extensible batch materialization engine.

- Feast SDK provides basic capabilities in fetching feature retrieval for model training and serving.

- A fairly extensive documentation at `https://docs.feast.dev` that includes tutorials and how-to-guides.

Opportunities

- Limited data quality capabilities out of the box. However, there is an upcoming initiative around data quality monitoring to help validate the data with user-curated set of rules.

- Real-time feature transformation and ingestion are a complex and would love to see Feast steps up in providing a general solution for the community.

Feature store is a common need for organizations that are looking to productionize their ML models and particularly at scale. Feast has the opportunity to become the de facto open source project for this need, similar to how Apache Spark has become the de facto open source project for distributed and scalable data computation. A good resource to explore and learn about Feast is at `https://feast.dev/`.

In-House

There is no shortage of presentations and best practices being shared about the in-house feature store or feature engineering infrastructure solutions from numerous companies like AirBnb, LinkedIn, Uber, Meta, Twitter, etc. The annual Feature Store Summit[15] is a great place to discover these resources. It is challenging to study and explore these in-house solutions in detail due to the limited availability of the design, source code, and examples. However, in April 2022, LinkedIn announced the open source of their feature store solution called "Feathr: a scalable, unified data and AI engineering platform for enterprise"[16] under the Apache 2.0 license and Feathr has joined the LF AI & Data Foundation.[17]

---

[15] Feature Store Summit, `www.featurestoresummit.com`

[16] David Stein, "Open sourcing Feathr – LinkedIn's feature store for productive machine learning," 2022, `https://engineering.linkedin.com/blog/2022/open-sourcing-feathr---linkedin-s-feature-store-for-productive-m`

[17] LF AI & Data, `https://lfaidata.foundation/`

This is very exciting news for the open source community because Feathr is designed to address the common needs in simplifying ML feature management, and it has been battle tested at LinkedIn for more than 6 years through numerous complex ML and real-time AI applications, such as Search, Feeds, and Ads, and their ML models are powered by thousands of features with petabytes of feature data.

According to this blog,[18] Feathr helps improve feature engineering velocity by reducing time to days from weeks to add and experiment with new features and has proven to perform up to 50% faster than custom feature processing pipelines.

The following sections will examine Feathr in detail with respect to its capabilities and architecture.

Overview

Before Feathr was created, teams at LinkedIn were responsible for building and maintaining their feature pipelines using the various in-house tools and libraries from the data infrastructure team. As the feature pipeline complexity increases, the maintenance effort also increases, which reduces the amount of time that those teams have to focus on more high value ML-related tasks. Feature reuse was not that easy due to a lack of standardization and centralization; therefore, it was challenging to reuse features across similar projects.

At the highest level, Feather provides an abstraction layer to standardize and simplify feature definition, transformation, serving, storage, and access from with ML workflows, as depicted in Figure 3-5. Let's discuss the provided abstractions from the two typical personas that are involved in the feature engineering: producer and consumer.

---

[18] "Hangfei Lin," "Feathr joins LF AI & Data Foundation," 2022, https://engineering.linkedin. com/blog/2022/feathr-joins-lf-ai-data-foundation

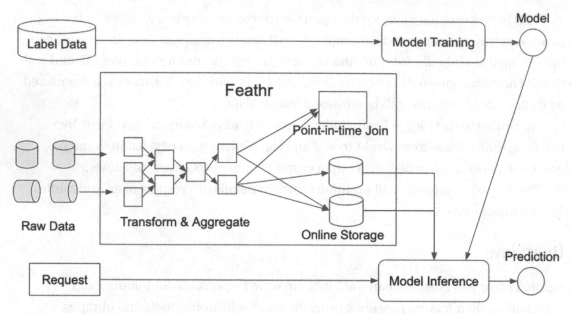

Figure 3-5. *Feathr feature store solution – adapted from the Architecture Overview[19]*

Producer

This persona is primarily responsible for defining, generating, and inserting features into the feature store by

- Define and register one or more features based on either raw dataset or pre-compute dataset from one or more data sources using Feathr Python APIs

- Leverage the various built-in transformations, aggregations, time windowing to enrich the feature values that are of various types, including vectors and tensors

- Leverage existing features to create new ones via the derived feature support

[19] David Stein, "Open sourcing Feathr – LinkedIn's feature store for productive machine learning," 2022, https://engineering.linkedin.com/blog/2022/open-sourcing-feathr---linkedin-s-feature-store-for-productive-m

Consumer

As the name suggests, this persona is mainly responsible for consuming the already registered features for either model training or model serving purpose and let Feathr figure the underlying means to provide the specified features.

- Compute and fetch historical values of features using feature names and join keys, and ensure they are done in point-in-time correct way.

- During model serving phase, materialize features onto online store, and make them available during online inference.

A few notable sophistications that Feathr provides:

- Rich-type system: In addition to the standard types in tabular data, Feathr also supports both embeddings and tensors.

- Rich support for complex transformations: Feathr has high performant built-in operators for time-based aggregation, sliding window joins, and all with point-in-time correctness.

- Scalable built-in optimizations: Native optimization like bloom filters, salted join, and built-in join plan optimizer.

- Feature sharing and reuse: There is a built-in feature registry for feature discoverability and an easy way to create derived features based on existing one.

One interesting factor is that the Microsoft Azure team complements the Feathr open source project by adding necessary improvements and support to make sure it is easy to integrate with cloud vendors, such as Azure.

Concepts

The abstractions mentioned above operate on a set of core concepts that Feathr provides and for its user to define. The core and common set of concepts among the different feature store infrastructures are quite similar; however, the concept names tend to vary due to the personal perspective and experience of their creators.

Similar to the ones in Feast, the concepts are structured in a hierarchy with three levels.

The lowest level consists of Sources, Features, Derived Features, and Anchors, and their metadata will be managed by the feature registry. A brief description of each one is provided as follows:

- Source: Represents the source data that the features are extracted from. Examples of the supported sources are distributed file systems like HDFS, S3, Azure Storage Blog, as well as streaming sources like Kafka. An optional preprocessing callback can be provided to perform transformation on each row.

- Feature: Represents a measurable property of an entity and typically has a unique name, key, type, and transformation logic to produce its feature value. The feature transformation logic can be a simple one, such as casting from one type to another, or can be as complex as a window aggregation. Feathr provides numerous useful built-in transformations and supports custom ones.

- Derived feature: A way to promote feature reuse by building features on top of existing features with certain transformations.

- Anchor: Designed to bring together one or more features that come from the same source.

The second level has only one concept, and it is referred to as Feature Query. This is designed to be used by the feature consumers for purposes such as exploration, analysis, or model training. This concept enables feature consumers to select a set of features to retrieve their feature values from the specified dataset, which is called observation data in Feathr.

The third level contains a single concept called Project, which represents the top-level namespace for managing one or more features. This is one way to manage features that belong to a particular ML use case, team, or organization.

Architecture

As a feature store, Feathr's architecture is quite similar to Feast's architecture in terms of the number of components, their role, and data flow, as depicted in Figure 3-6.

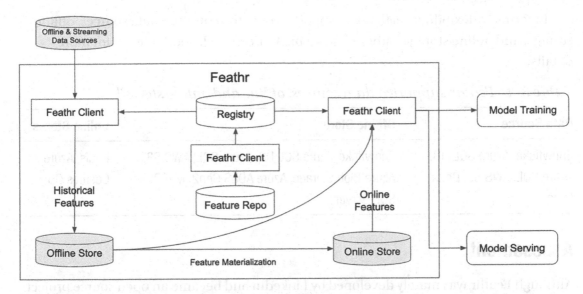

Figure 3-6. *Feathr architecture – adapted from the Feathr documentation on GitHub[20]*

However, there are a few significant differences that are not obvious from Figure 3-6 and are worth calling out.

One key and distinguishing feature Feathr provides is the ability to perform feature transformation at scale during the feature access or generation step (aka feature materialization). This starts with the first class support in capturing the feature transformation logic in the feature definition, which then can be reproduced and version-controlled and promotes feature reuse due to having more visibility into how those features are computed. Coupled with this feature is the high performance built-in operators to perform various complex transformations like point-in-time joins, time-aware sliding window aggregation, etc. Without this, data scientists would spend many hours of their precious time on data wrangling, performing complex joins in an optimal manner. Without a doubt, this capability is one of the reasons data scientists at LinkedIn can productionalize their features in hours, which used to take weeks.

A key design decision the Feathr team made is to leverage Spark engine as the compute engine for performing feature generation and aggregation, and this has helped them to achieve the flexibility and scalability goals.

---

[20] "Feathr Registry and Feathr UI," https://feathr-ai.github.io/feathr/concepts/feature-registry.html

In terms of flexibility, Feathr can support most of the common data sources, offline storage, and online storage, although it is a bit Azure biased. See Table 3-2 for more details.

Table 3-2. *Feathr-supported data sources, offline and online stores*[21]

Data Source	Offline Store	Online Stores
Snowflake, Azure SQL DB, Azure SQL, AWS S3, Delta Lake	Snowflake, Azure SQL DB, Azure SQL, AWS S3, Azure Blob Storage, Azure ADLS Gen2, MySQL, SQL Server	Redis, Azure Cosmos DB

Assessment

Although Feathr was mainly developed by LinkedIn and became an open source project in 2022, its capabilities and features are quite impressive. It meets all the standard feature store requirements and more. From the maturity perspective, Feathr has been battle-tested at LinkedIn over 6 years and serving thousands of features in production.

The Azure team and LinkedIn team have been collaborating closely to build native integration between Feathr and Azure. This means it will only get better and more matured as the adoption increases from Azure customers.

Strengths

- A full-fledged feature store solution with feature definition, feature registry, UI, and SDK for feature generation, feature serving and both offline and online support

- Native support for embeddings and promotes feature reuse via derived feature support

- A built-in support for feature transformation with an extensive built-in operators and extensibility via UDF from Spark SQL or user-defined ones

---

[21] "Feathr Cloud Integration," https://github.com/feathr-ai/feathr#cloud-integrations

- A built-in support for point-in-time joins to prevent feature leakage

- Native integration with Spark compute engine for scalable feature generation and join optimizations

- Provide role-based access to feature registry via Microsoft Purview integration

- A fairly decent feature management UI that includes feature lineage

- A fairly extensive documentation at `https://feathr-ai.github.io/feathr/` that includes tutorials and how-to-guides

Opportunities

- No explicit data quality integration.

- No feature monitoring support at the moment; however, it is on the roadmap.

Time will tell Feathr's viability and what the community adoption will be like. As a sign of committing to open source, both the leadership at LinkedIn and Microsoft shared their support for Feathr to join the Linux Foundation AI & Data as a new sandbox project[22] at the end of 2022. In terms of the capabilities Feathr provides, they are quite extensive and impressive when compared to other open source solutions.

Vendor Solutions

It is reasonable to say 2022 had been the year of the feature store. Both the challenges and needs in the feature engineering infrastructure area are well understood and recognized; therefore, it is not a big surprise to see numerous vendor solutions introduced in 2022 from both cloud vendors and cloud-native vendors.

---

[22] "Erin Thacker," "Feathr Joins LF AI & Data as New Sandbox Project," 2022, `https://lfaidata.foundation/blog/2022/09/12/feathr-joins-lf-ai-data-as-new-sandbox-project/`

The notable ones are Vertex AI feature store, Sagemaker feature store, Databricks feature store, and Tecton feature platform. It is reasonable to expect most if not all these solutions check the boxes of the common requirements of a feature store:

- Feature reuse and discovery through feature meta and registry

- Offline and online store support to ensure consistency and no training-serving skew

- Point-in-time feature retrieval

- Feature lineage tracking

- Feature monitoring

Among the solutions mentioned above, the one from Tecton has branded itself not as a feature store, but rather a feature platform, which is the closest to the feature engineering infrastructure described in this chapter. This section below will get into the specifics of the Tecton feature platform.

Overview

The company Tecton was founded by a group of engineers that were helping with building the Uber Michelangelo ML platform, which was instrumental in enabling Uber to scale to 1000s of models in production, supporting a broad range of interesting ride sharing and marketplace-related use cases from real-time pricing, ETA prediction, and fraud detection. An important contributor to Michelangelo ML platform's success is the feature store, enabling feature creation and serving quickly and reliably in production.

The initial feature store commercial product version was unveiled in 2020, and it has evolved and expanded into a full-fledged feature platform built to orchestrate lifecycle of features, from transformation to online serving, with enterprise capabilities.

The discussion below is not meant as an endorsement for Tecton product, rather it is meant to illustrate the various capabilities and features of what an ideal feature engineering infrastructure looks like.

Concepts

It is not a big surprise that most concepts and their hierarchy structure in Tecton feature platforms are quite similar to the ones in the other feature stores.

The lowest level consists of Data Sources, Entities, Feature Views, and Feature Tables. A brief description of each one is provided here:

- Data source: Represents the data source to read the raw data from. There are two types: batch and stream. Examples of batch data sources include a file on S3, a table in Hive, a query or table in AWS Redshift, or a query or table in Snowflake. The supported stream data sources include Kafka and Kinesis.

- Entity: Typically represents a domain object that has one or more primary keys and a set of related features. Examples include Customer, Product, and Order.

- Feature view: A core abstraction to define one or more related features as a view on registered data sources or other feature views. The feature specification and metadata are specified in Python. There is where the configuration about materialization is specified, such as orchestration and either offline or online destination.

- Feature table: Enables the ingestion of pre-transformed features into Tecton.

The second level has only one concept and it is referred to as Feature Service. It is designed to be used by feature consumers for purposes such as batch lookups of feature values during model training or offline prediction, or low-latency requests for individual feature sets during online predictions. A feature service references a set of features from one or more feature views. As the name implies, a feature service provides a set of capabilities behind an actual service that includes

- A REST endpoint to fetch feature values at the time of prediction

- A very simple method one-line method call to quickly constructing training data based on the provided timestamps and labels

- The ability to continuously monitor the online requests and feature vector responses via logging those information for auditing, analysis, and training dataset generation

A good practice is to pair each model deployed in production with one feature service, which serves features to the model.

The third level contains a single concept called workspace, which represents the top-level namespace for managing the various concepts first and second level. All the operations related feature definition, materialization, feature access, and more will need to start with selecting which workspace to work with. This is similar to the concept of selecting a database before working with their tables, table schemas, and the associated data.

Architecture

According to the Tecton documentation website,[23] their feature platform is a fully managed feature platform to orchestrate the complete lifecycle of features, from transformation to online serving, with enterprise-grade SLAs. One of the notable aspects that distinguishes a feature platform from a feature store is the orchestration of existing data infrastructure to continuously transform, store, and serve data for operational machine learning applications. At the high level, its architecture, depicted in Figure 3-7, is not that drastically different from the solutions mentioned previously.

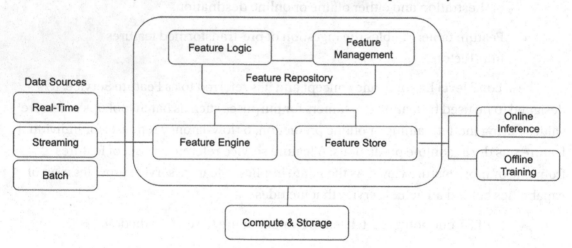

Figure 3-7. *Tecton feature platform architecture. Adapted from an image from Tecton Concepts documentation*[24]

---

[23] "What is Tecton?," https://docs.tecton.ai/docs/introduction
[24] "Tecton Concepts," https://docs.tecton.ai/docs/introduction/tecton-concepts

The Tecton feature platform was intentionally designed to not replace existing data infrastructure, but rather enable them for operational machine learning applications by connecting well-established batch and streaming data sources, leveraging industry accepted compute engines, and supporting commonly adopted storage infrastructure.

The major components in Tecton feature platform are feature repository, feature engine, and feature store. The bullet points below briefly examine their capabilities:

- Feature repository: Feature producers define feature specification and metadata using a Pythonic declaration interface in Python file. Besides the standard details like name, description, and transformation logic, feature producers can specify a configuration about how often the feature should be computed. Once the features are defined, they are available in a centralized repository for others to discover and reused.

- Feature engine: Responsible for orchestration of the feature pipelines to materialize the computed feature values and then publishing them to the feature store. In addition, it is responsible for interfacing with the underlying compute engines to execute the specified feature transformation logic.

- Feature store: Responsible for managing both the offline and online store, as well as serving features via the abstraction and APIs provided SDK. For online feature serving, all requests go through a feature service.

One important capability that is not obvious from the architecture diagram is the monitoring capabilities. Out of the box, Tecton feature platform provides data quality and operational monitoring. The former is about tracking incoming data distribution and quality. The latter is about monitoring feature staleness, storage health and efficiency, and the health of feature pipelines. The monitoring capabilities play a critical role in ensuring and maintaining the model performance continues once the ML models are deployed to production.

In terms of data sources and sinks, the Tecton feature platform can support most of the common data sources, offline storage, and online storage. See Table 3-3 for more details.

Table 3-3. *Tecton feature platform-supported data sources, offline, and online stores*

Data Source	Offline Store	Online Stores
Hive, S3, Snowflake Databricks, Redshift, Kinesis, Kafka	Hive, Glue, S3, Redshift	Redis, DynamoDB

Assessment

Given it is an enterprise and commercial vendor solution, there is a greater expectation that the Tecton feature platform provides a much more comprehensive set of features in the feature engineering infrastructure than the ones from Feast and Feathr. The provided capabilities are a super set of the ones from the other case studies; therefore, the strength and opportunities sections below will only call out the notable ones and will not repeat the obvious ones.

Strengths

- The ability to orchestrate feature pipelines via configurations simplifies the need to create and manage orchestration or workflow system as a separate piece of infrastructure.

- The provided monitoring of data distribution, data quality, and operational aspect is critical to productionalizing ML at scales and to protect the downside when model performance starts to degrade or unexpected data-related issues pop up.

- The ability to define the transformation logic outside of the feature definition and have them discoverable in the UI encourages reusability. This is very valuable when multiple teams work on similar ML use cases and there is a reasonable amount of feature overlapping.

Opportunities

- The feature name is implicitly derived from the feature transformation function name, as such it is not obvious at first. Promoting feature name as a first class feature metadata will facility feature sharing and feature reuse.

Summary

It is a well-known fact from the ML practitioner community that data scientists spend a significant amount of their time in the feature engineering step of the ML development process. The feature engineering infrastructure aims to not only reduce the amount of time to generate features, but also increase collaboration and best practices among the teams that are applying ML. The benefits are clear, and it is vital for organizations to get this right, especially for online ML use cases, as they are scaling up their ML investments.

It is a complex piece of infrastructure and has a large set of dependencies on other infrastructures, such as data, storage, and compute. The core set of components in this infrastructure includes

- Feature specification and definition
- Feature registry
- Feature pipeline orchestration
- Feature store
- Feature serving
- Feature monitoring

In terms of the possible solutions, this chapter examined a solution from each source: open source, in-house, and commercial vendor. Their strengths, opportunities, and assessments are provided. The build vs. buy decision is never easy, and there are many contributing factors into the final decision. Hopefully the provided high-level assessments are useful as a starting point.

CHAPTER 4

Model Training Infrastructure

In the ML development process, the phase that follows feature engineering is known as model training. This crucial phase involves selecting and deciding an ML algorithm from the pool of diverse options, and training it using the selected features. The objective is to train the ML algorithm to learn patterns within these features so it reasonably can make accurate predictions on new and unseen data. The model training pipeline encompasses several key steps: ML algorithm selection, model training, hyperparameter tuning, and model evaluation. In a typical ML project, it is necessary to iterate through these steps multiple times to attain a high-performing and generalized model that meets the specified performance metrics and business requirements.

The algorithm selection involves choosing an ML algorithm for the specific task at hand, whether it is a regression or classification task. This step is often described as both an art and a science. The art part comes into play when trying to understand the nuances of the business problem at hand, the ML task, and the available data characteristics, and then figure out which ML algorithms might be best suited for it. It takes years of experience to build good intuition as well as creativity in order to be effective in this step. The science part is rooted in a systematic and data-driven approach. This involves leveraging statistical analysis to analyze the available data, identify potential biases, and evaluate the model performance to narrow down a potential set of adjustments for the next experiment. In summary, the art part is about leveraging intuition and experience to make educated guesses, while the science part performs systematic experimentation, analysis, and adhering to the established best practices.

The hyperparameter tuning step involves determining which internal knobs of the algorithm to adjust to improve model performance. Similar to performing a chemistry experiment, it requires running multiple experiments with different parameter values and evaluating the model's performance after each experiment against the set of metrics defined at the beginning of the ML project. The goal of this step is to find an optimal set of hyperparameters that result in the best performance on the specific ML task.

Note   Model parameters vs. hyperparameters

In the context of model training and tuning, model parameters and hyperparameters are two important concepts, and they serve different purposes.

Model parameters are learned by the ML algorithm during the training process. They are not something data scientists provide or determine. Examples of model parameters are the coefficient and weights that the ML algorithms adjust during training to minimize the difference between the actual target values and the predicted output.

Hyperparameters are the tuning knobs that data scientists set before the training process begins to influence the how or the speed the ML algorithm learns. Examples of hyperparameters include the learning rate, batch size, the number of hidden layers, and more.

At the high level, the model evaluation phase is designed to assess the performance and generalization capabilities of the trained model. To achieve these goals, first the appropriate evaluation metrics are chosen based on the specific task, then collect those metrics during the algorithm training step, and finally assess how well the model generalizes to unseen data. The common metrics include accuracy, precision, recall, F1-score, mean squared error (MSE), and area under the receiver operating characteristic curve (AUC-ROC).

The ultimate objective of the model training phase is to produce a generalized and high-performance model that aligns with the business requirements. This alignment could be about enhancing customer experience, increasing certain business metrics, or reducing operational cost. To successfully support these goals, it will require numerous pieces of infrastructure to make it easy and to speed up the iteration, and to support various software engineering needs, such as compute resources and orchestration, and distributed model training for complex deep learning models with a large volume of features.

Overview

Model training infrastructure is the second and crucial pillar of the overall machine learning infrastructure. Its primary objective is to provide an array of resources, tools, and infrastructure for all the activities related to training machine learning models. This includes an integrated development environment, experiment tracking, computational resources for large-scale model training and offline evaluation, visualization tools, and more.

Each step in the model training phase is unique and has its own distinct requirements. Consequently, the infrastructure designed to support these steps consists of a diverse array of components. These components must be connected in a cohesive and integrated manner, ensuring a delightful and seamless experience that promotes fast iteration and experimentation. Moreover, special attention is needed when supporting large-scale ML projects.

These are the common benefits that model training infrastructure provides:

- Efficiency

  - As described above, the model training process involves multiple steps. Streamlining this process to make it more efficient and time-effective would allow data scientists to dedicate more time to ML specific tasks, such as model tuning and experimentation.

- Collaboration

  - Medium to high-complex ML projects typically require collaboration among multiple data scientists. This collaboration can be promoted and supported with easy access to shared resources, version-controlled model artifacts, and other collaboration tools. This can lead to better ML models and improved outcomes for businesses.

- Experiment management

  - One of the key steps in improving model performance is experimenting with different combinations of model hyperparameters. Providing experiment management and tracking tools to make it easy to organize, track, analyze, and reproduce experiments. This can speed up the process of comparing different model performance and identifying the most effective approaches.

- Scalability

    - Complex ML use cases often utilize deep learning techniques to
      learn from a vast amount of data. This requires training models
      in a distributed manner and leveraging powerful hardware.
      The scalability of the model training infrastructure enables
      faster training and experimentation, which in turn, facilitate the
      discovery of new approaches to enhance business outcomes.

The following sections will delve into the details of the model training infrastructure
architecture. This will include an exploration of the interconnections between the
various components, along with insights into the specifics of each component.

High-Level Architecture

An effective model training infrastructure should provide a cohesive toolkit capable of
meeting the needs of a diverse set of steps and accommodating the scalability demands
of the ML workload scale during the model training phase. For companies in the initial
stages of adopting machine learning, their model training infrastructure might not
require the same level of sophistication as those that have been extensively utilizing
machine learning over many years and at a larger scale.

Fortunately, the tools and infrastructure provided by the open source communities
and MLOps vendors have steadily matured over the last few years. As a result, companies
now have the flexibility to combine and customize the necessary solutions to the
specific needs.

From the engineering's perspective, the model training infrastructure consists of the
following components:

- Model development environment provides an easy to use and
  interactive model development environment for data scientists to
  quickly develop their ML models.

- Experiment tracking helps with record and manage the various
  metadata while data scientists are running various model
  development experiments. Data scientist then can easily visualize the
  experiment results or compare them across the experiment to better
  understand the model performance.

- Model training provides facilities to manage model training pipelines to promote reproducibility, and access to compute resources to train ML models in both ad hoc and scheduled basis.

- Model store acts a centralized repository to manage the model metadata, artifacts, and model lifecycle.

Figure 4-1 depicts the high-level components and their interactions. The input is the features that are stored on the feature store, and the output is the trained models that are ready for offline inference or online inference. The middle box encompasses the tools, frameworks, and infrastructure to support the main aspects of the model training process.

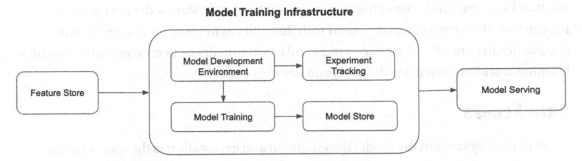

Figure 4-1. *Model training infrastructure high-level architecture*

Unlike the feature engineering infrastructure, there isn't a readily available end-to-end vendor solution that can be simply adopted. However, there are open source and vendor solutions for each of the components listed above. We will explore some of those options in the case studies section.

The upcoming sections will provide a brief overview of the distinct capabilities of each component while also highlighting common approaches from the industry and open source communities.

Model Development Environment

During the model development and exploration stage, a highly interactive, collaborative, and flexible development environment is an indispensable tool for data scientists. This environment should offer rapid feedback and visualization capabilities, enabling them to quickly perform data analysis, visualizing the results, iterate on the development of ML models, and communicate their findings with their peers.

The de facto and widely popular development environment in the ML community is called Jupyter web-based interactive development environment. According to their website,[1] Jupyter is an open source software and branded as "free software, open standards, and web services for interactive computing across all programming languages.

Their latest notebook interface is called JupyterLab, which offers a flexible and powerful user interface that combines code, documentation, data, rich visualizations, graphs and figures, and interactive controls. It is quite easy to install and get JupyterLab running on a local machine with just a few commands.

For teams or companies that would like to provide a distributed and shared Jupyter Notebook environment for their data scientists to use without having to install software on their local machine, they can leverage the multi-user version of the notebook called JupyterHub. This project was designed with flexibility in mind to enable teams and companies to control access to resources and environments or to customize the machine learning workflow and using the tools and libraries of choice.

Data Access

A few of the key steps in the model development and exploration is the data analysis and visualization. To do so, data scientists need to have access to the centrally managed data repository, often known as data warehouse or data lake, that contains a diverse set of datasets and some of them might be large in volume, as well as some of the shared features that are generally developed and used in various ML-related projects. As such, the development environment should provide easy and safe access to the data repository. Without this access, the model development and exploration will be hampered and slow.

Compute Resource Access

Once data scientists have access to the needed dataset or available features, they will start analyzing them and evaluating whether they are suitable for the ML task at hand. To perform medium- to large-scale data analysis, they will need access to compute resources beyond their laptop so those data crunching needs will be completed in a short amount of time.

---

[1] Project Jupyter, https://jupyter.org/

This is where distributed data computation engines come into the picture. Examples of these engines are Apache Spark, Dask, and Ray.

Note   Spark vs. Dask vs. Ray

All three are popular distributed computing frameworks that are widely adopted for diverse data processing needs. Spark is the most mature one and has a large ecosystem. It provides a robust unified platform for large-scale data processing with high-level APIs in multiple languages, such as Java, Scala, and Python. Dask excels in parallel computing and seamlessly integrating with popular Python libraries, making it well suited for scalable and complex data computations. Ray, on the other hand, is a compute framework to enable efficient distributed execution of Python and AI workloads, boasting a simple programming model and automatic parallelization.

Ultimately, the choice among these frameworks depends on the specific demands of tasks: Spark for general big data processing, Dask for Python-centric parallelism, and Ray for distributed and ML workloads.

Similar to the above section, the development environment should provide an easy way to access the needed compute resources to perform data analysis at scale. As the compute resource access increases with frequency and scale, it is advisable to invest in cost attribution and cost management capabilities, such as shutting down clusters when they are idle after a certain amount of time or inactivity.

Model Development Experience

While training large neural network or tree-based models, data scientists would like to have visibility and insights into the model training process to better understand the progress by observing key metrics such as loss and accuracy as they update during each iteration. By understanding how well their model is learning, data scientists can stop the model training early and wouldn't have to waste multiple hours of their precious time if the learning is not going the right direction, whether due to the overfitting or convergence issues.

A few specific tools that are tremendously valuable to improve the model development experience are the TensorBoard from the TensorFlow project and the model agnostic methods to understand model interpretability, such as LIME, SHAP, and ICE Plots.

TensorBoard is a widely popular and powerful visualization tool for machine learning experimentation provided by TensorFlow, an open source machine learning project developed by Google. Many popular machine learning libraries now provide integration with TensorBoard, such as Pytorch, XGBoost, Ray Train, HuggingFace, and more.

Note   Model interpretability

This refers to the ability to understand and explain how an ML model makes predictions. By providing insights into the factors influencing predictions, model interpretability aids in debugging, refining, and maintaining models in real-world applications.

According to the TensorBoard document,[2] here are some of the key features it provides:

- Real-time tracking and visualization metrics such as loss, accuracy, precision, and recall. This is useful to understand how one's model is learning.

- Model graph visualization displays a graph of the model's computation graph. This is useful for understanding the architecture and the data flow.

- Histograms and distributions visualization displays the distribution of weights and biases in the model. This is useful for understanding issues like overfitting.

---

[2] TensorBoard from TensorFlow project, `www.tensorflow.org/tensorboard`

- Embeddings visualization can project high-dimensional data into lower-dimensional spaces. This is useful for understanding the relationships between data points.

- Profiler is a profiling tool to help with performance analysis of the model to identify potential bottlenecks.

Tools like TensorBoard and similar ones will significantly improve the experience of machine learning development. As a result, data scientists will be more productive at model development and experimentation. Figure 4-2 depicts the visualization of the accuracy and loss metrics at training and validation steps.

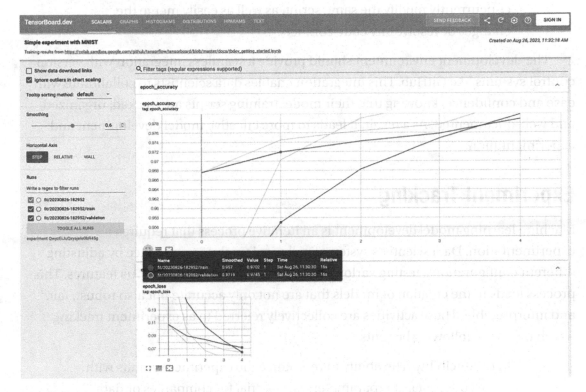

Figure 4-2. *Tensorboard visualization of a few metric examples*

Reproducibility

One of the main principles in MLOps is reproducibility. Once the model development and exploration is settled down, the next logical step is to version control the model training code. This serves the following purposes:

131

- Traceability: Version control ensures changes made to the script can be tracked back. This is useful when other data scientists would like to reproduce the model.

- Debugging and troubleshooting: When unexpected issues come up after a model is deployed to production, version control allows data scientists to revert to a known working state or to identify the changes that caused the issues.

- Collaboration: In some cases, there might be multiple data scientists working on the same model. Version control enables them to concurrently modify the same script, as well as easily merge the changes or resolve any code conflicts.

The development environment should provide a seamless integration with version control systems like GitHub. This integration enables data scientists to collaborate with ease and confidence, knowing that their model training scripts are tracked, organized, and well-documented. As a result, it leads to more effective model development and experimentation.

Experiment Tracking

Machine learning model development is an iterative process that requires experimentation. Data scientists systematically explore the search space by adjusting different configurations, testing various algorithms, and selecting various features. This process leads to the creation of models that are not only accurate but also robust, fair, and interpretable. These activities are collectively referred to as experiment tracking, which offers the following benefits:

- Reproducibility: The ability to reproduce the experiment results with same code, data, and parameters is essential for companies or data scientist teams to verify results and diagnosing issues.

- Collaboration: Tracking experiments allow data scientists to easily share insights with each other and to collaborate on potential ideas to improve model performance or overcoming challenges.

- Hyperparameter optimization: In addition to tracking different combinations of hyperparameters and configurations, experiment tracking makes it easier to identify which setting lead to the best model performance.

- Decision-making: By analyzing past experiments to understand the trends, patterns, and failure, these insights guide data scientists in making better future choices.

Historically, data scientists use a spreadsheet to track the experiment inputs and experiment results. However, this is both tedious, error prone, not scalable, and ineffective to analyze and compare the results across multiple experiments.

As the MLOps discipline starts to mature, both the open source community and vendors recognize the need for a robust solution to experiment tracking. Before diving into the solutions, let's outline the common requirements associated with experiment tracking:

- Metadata management: The ability to track the various and diverse metadata associated with each experiment (e.g., author, environment, dependencies).

- Data management: The ability to store large volumes of both structured and unstructured data generated during experiment efficiently. This includes the input data, model outputs, metrics, and various artifacts.

- Scalability: As the number of experiments increases, managing and tracking them becomes more resource-intensive.

- Analysis and visualization: Effective analysis and rich visualization of diverse metrics across multiple experiments will assist data scientists in making informed decisions for their next steps in the experimentation process.

At the time of this writing, there are numerous experiment tracking solutions (from open source and vendor) available for us to choose from. All these solutions provide an easy way to integrate experiment tracking into the model training code via Python library, and some of them support automatic logging of metrics, parameters, and models and lineage information without the need for explicit log statements. One popular and widely adopted experiment tracking solution from the open source community is

MLflow[3], which is branded as the platform for managing machine learning lifecycle, including experimentation, reproducibility, deployment, and a central model registry. This platform consists of four components:

- Tracking: Record, query, and visualize experiments through APIs and UI.

- Projects: Format for packaging data science code to reproduce runs on any platform.

- Models: Standard format for packaging machine learning models that can be used in a variety of downstream tools, such as deployment.

- Model registry: Central repository for storing, annotating, discovering, and managing the lifecycle of models.

Figure 4-3 shows the Experiment homepage of MLFlow 2.7.1 version. The table format displays a list of experiments with their associated metadata. From here you can easily compare the results of two or more experiments by clicking the associated checkboxes in the first column and then the Compare button.

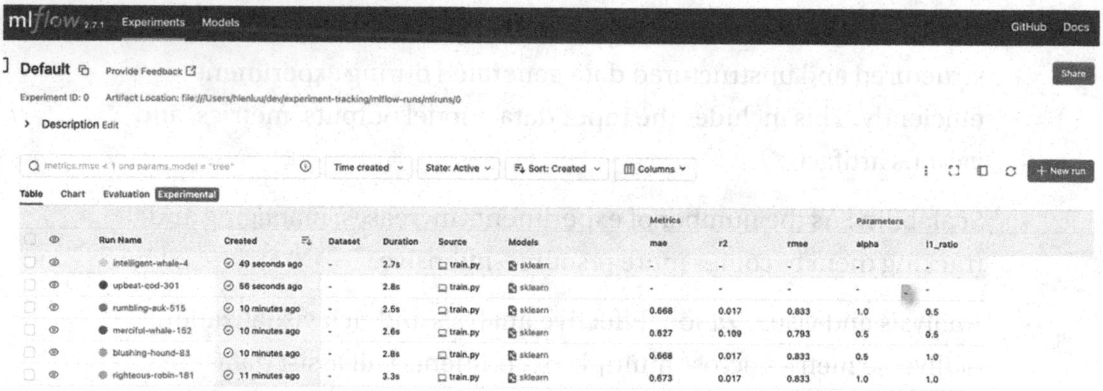

Figure 4-3. *MLFlow experiment tracking*

To see the details of an experiment like its metadata, parameters, metrics, and artifacts, click on its name under the "Run Name" column, and you will see something like in Figure 4-4.

---

[3] MLflow, an open source platform for machine learning lifecycle, https://mlflow.org/

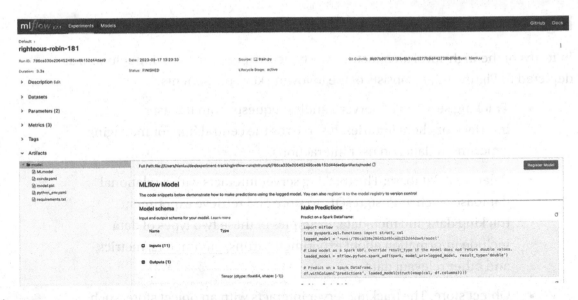

Figure 4-4. *MLFlow experiment details*

Considering the maturity of the various experiment tracking solutions available today, it makes a lot of sense to either adopt one of the open source solutions or opt for one of the vendor solutions. These solutions have evolved significantly and offer a rich set of features, and often provide integration with popular machine learning libraries. By leveraging these mature solutions, organizations can benefit from established best practices, ongoing support, and a wealth of resources from the open source or user community.

Most of the available solutions provide a pretty comparable set of capabilities to meet the experiment tracking needs, including the ability to log experiment information and metadata, an easy and intuitive way to visualize and experiment metrics and compare them across multiple experiments, a way to organize and search experiments and metadata, and a seamless integration with the various popular machine learning libraries.

This experiment tracking comparison blog[4] provides comprehensive and valuable insights into the 15 popular experiment tracking and management tools as of December 2021. It includes both vendor and open source solutions. Most vendor solutions are hosted solutions running in your favorite cloud providers, and they do provide a free tier for easy evaluation and exploration.

---

[4] 15 Best Tools for ML Experiment Tracking and Management, `https://neptune.ai/blog/best-ml-experiment-tracking-tools`

High-Level Architecture

In terms of the high-level architecture of a typical experiment tracking solution, as depicted in Figure 4-5, it consists of the following key components:

- Tracking server: The server handles requests from the user interface or client libraries. It serves as the central hub for managing experiment data and user interactions.

- Relational database: The tracking server interacts with a relational database, such as PostgresQL, to store and retrieve experiment tracking data and metadata. Examples of these two types of data include information about experiments, runs, parameters, metrics, and other relevant data.

- Object store: The tracking server interacts with an object store, such as AWS S3 or Google Cloud Storage, to store and retrieve experiment artifacts. These artifacts include model binary, images, and other files generated during the experimentation process.

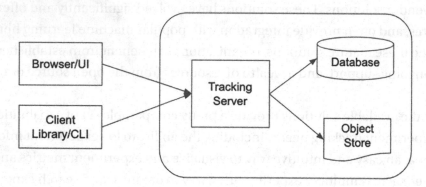

Figure 4-5. *Experiment tracking system architecture*

Undoubtedly, the experiment tracking component is a fundamental component of model training infrastructure. From an infrastructure perspective, seamless integration of experiment tracking during the model training development phase is crucial.

Model Training Pipelines

Once the model exploration and experimentation activities, which are often ad hoc, have reached a point where data scientists have developed a model that meets the target performance metrics, the next step is to package the end-to-end model training steps into a model training pipeline to produce the machine learning model. The pipeline serves as a workflow that connects a sequence of interrelated tasks, such as data preprocessing, the model training function, model evaluation, model deployment, and more. Model training pipelines enable data scientists to deploy their models to production easily and more importantly frequently.

Following the software engineering best practices, model training pipelines are treated as code that are reviewed by peers to ensure coding best practices are followed, such as documentation, modular code for understandability and maintainability, and versioned control.

Some of the benefits model training pipelines provide are

- Efficiency: Pipelines codify the sequence of steps, making it easy to automate repetitive tasks, thereby saving time and reducing manual intervention.

- Modularity: Structuring the sequence of steps in a well-defined manner enables changes without disrupting them.

- Consistency: Once the steps in the pipeline are codified, this ensures they remain consistent and make it easy to transition from development to production environments.

- Ease of experimentation: Well-defined and structured pipelines make it very easy to run experiments and compare them. This can help to increase the number of model training iterations, which can lead to better models.

- Efficiency: Data scientists typically run the pipelines multiple times with only a small change to one or two of the steps; the outputs of the unchanged steps from the previous training run can be reused to speed up the overall time model training time. A good example of this scenario is when the change is applied only to the hyperparameter setting, and therefore the data preprocessing steps from previous training can be reused.

Orchestration

Machine learning pipelines are about expressing the various machine learning tasks or steps that need to happen and their sequence, whether that is sequential or parallel. Building them and operating them can be time consuming for data scientists. This is where a modern machine learning friendly and data-driven orchestration system comes into the picture to help accelerate the speed of improving ML model performance by making it easy to develop, schedule, monitor, and manage the pipelines.

Fortunately, the maturity of orchestration solutions has increased dramatically in the last few years to meet the demanding needs of the modern data and machine learning workload, both in terms of data volume, data pipeline complexity, and the scale machine learning model training. In addition, the number of options also increased significantly from a handful to more than a dozen. For example, this MLOps blog[5] contains a fairly detailed comparison matrix of thirteen orchestration tools across various important dimensions. The majority of the options listed on this blog are open source projects, and the promising and feature-rich ones are the commercial open source projects, which represents a win-win model for both the company and the open source community.

Machine learning pipelines typically consist of two types of steps: data manipulation steps and machine learning steps. Each step type has their own specific needs, and in an ideal case, we would like to adopt an orchestration tool that can support all of their needs.

Some of the specific needs the data steps have are

- Multiple data sources and sinks: The feature engineering steps might need to read data from various data sources, and then the output will be persisted in one of the commonly used data sinks. The ability to read data from various data sources in various data formats is very important.

- Data processing engine: The feature computation is typically executed by data processing engines like Spark or Snowflake. The ability to interact and leverage the various data processing engines is crucial.

---

[5] Best Machine Learning Workflow and Pipeline Orchestration Tools, 2023, https://neptune.ai/blog/best-workflow-and-pipeline-orchestration-tools

- Data as a first class citizen: Datasets or tables have their own specific properties, such as schema, ownership, data quality checks, partitions, and SLAs. An orchestrator that treats these properties as first class constructs will make it very easy for data scientists to get familiar with.

Some of the specific needs the machine learning steps are

- Library version flexibility: Each model training script potentially uses a different set of libraries or different versions of the same library. Supporting the library version is a must for machine learning pipelines.

- Compute resource management: Large machine learning models will typically require training with GPUs in order to speed up the training time. Built-in support for diverse compute resource allocation will greatly simplify the machine learning pipeline development.

- Experiment tracking: Native integration with experiment tracking tools for tracking and logging training experiment and broader MLOps tools like model registry will be extremely valuable to model training pipelines.

One common theme across numerous modern orchestration tools is the native integration with Kubernetes, which gives them numerous benefits, such as isolation, scalability, and reliability.

- Isolation: Each machine learning pipeline is running in its own separate Kubernetes pod, and this allows each pipeline the freedom to use its own libraries and version and helps prevent library version conflicts.

- Scalability: Kubernetes is widely used and battle tested to manage large clusters of resources to support the workload demand of machine learning pipelines.

- Reliability: Kubernetes provides a number of features that can help machine learning pipelines reliable, such as automatic restarts and self-healing.

Note   Kubernetes overview

Kubernetes is a widely adopted open source container orchestration platform. Most modern containerized applications are deployed on Kubernetes clusters nowadays. Kubernetes helps with automating deployment, scaling, and management of containerized applications. A machine learning pipeline is an example of a containerized application.

Orchestration Programming Style

In the context of machine learning pipelines, data scientists are the ones that will be tasked with putting together or building these pipelines. Their choice of pipeline development style might vary due to their background, personal interest, and more. However, it's a well-established fact that the Python programming language is the lifeblood of most data scientists. It is used for a wide variety of data science tasks, including data cleaning, data analysis, model training, and evaluation. In addition, there is a very healthy ecosystem of data science libraries that are written in Python. As such, data scientists will feel more at home if the style is more pythonic.

In general, the orchestration tools provide two common ways of building workflows, which are the same as machine learning pipelines. The first approach involves expressing the workflow structure, including steps, sequence, and dependencies, in a YAML file. Figure 4-6 shows the hello-world example[6] of Argo workflow in YAML. The second way to express the workflow structure is adding workflow-specific decorators inside the Python code functions that contain the actual step logic. Figure 4-7 shows another hello-world[7] example from Flyte orchestration tool.

---

[6] Argo hello-world example, https://github.com/argoproj/argo-workflows/blob/master/examples/hello-world.yaml

[7] Flyte hello-world example, https://github.com/flyteorg/flytesnacks/blob/master/examples/basics/basics/hello_world.py

```
apiVersion: argoproj.io/v1alpha1
kind: Workflow
metadata:
  generateName: hello-world-
  labels:
    workflows.argoproj.io/archive-strategy: "false"
  annotations:
    workflows.argoproj.io/description: |
    This is a simple hello world example.
spec:
  entrypoint: whalesay
  templates:
  - name: whalesay
    container:
      image: docker/whalesay:latest
      command: [cowsay]
      args: ["hello world"]
```

Figure 4-6. *Workflow in YAML file*

```
from flytekit import task, workflow

@task
def say_hello() -> str:
    return "hello world"

@workflow
def my_wf() -> str:
    res = say_hello()
    return res


if __name__ == "__main__":
    print(f"Running my_wf() {my_wf()}")
```

Figure 4-7. *Workflow using Decorators in Python file*

The two styles of expressing workflow structure mentioned above are very distinct, not only in the format, but more importantly, in their mental model and friendliness from the data scientist's perspective. Combining the expression of workflow structure with the actual logic in a Python file would likely be more warmly received by data scientists when constructing machine learning pipelines.

There is a good selection of well-designed and mature open source orchestration tools for us to choose from, including Argo, Dagster, Flyte, Metaflow, Prefect, and more. The process of adopting one to power machine learning pipelines could greatly benefit from establishing a set of evaluation criteria. Several factors to consider include ensuring robust isolation to prevent library conflicts, assessing Kubernetes support for large complex pipelines, and giving priority to data scientist-friendliness.

Continuous Model Training

Machine learning models are made up of data, algorithms, and code, as depicted in Figure 4-8. Among the three components, data is the one that changes more rapidly, especially in the consumer Internet market. If the deployed models in production are not retrained with updated or changing data due to changes in the real world, the model prediction will gradually stop reflecting the new reality. Consequently, its performance will degrade when compared to the expected performance.

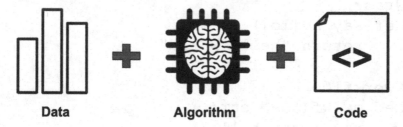

Data   **Algorithm**   **Code**

Figure 4-8. *The core composition of a machine learning model*

One of the types of data changes is known as "data drift." Data drift occurs when a statistical property of the production data used for model prediction, such as the distribution, changes or deviates from the baseline data used to train the model.

During the COVID-19 pandemic, there was a rapid and dramatic shift in customer behavior that caused data drift. Here are a few fun and easy ways to understand examples:

- Online shopping: During the pandemic, the way people shopped online changed, which affected the data used to predict shopping behavior.

- Air travel: The pandemic caused a big drop in air travel, which affected the data used to predict flight bookings and passenger traffic.

Maintaining model performance once they are deployed to production requires the ability to retrain and redeploy them in an easy, automated, and safe manner. In an academic sense, it is a reasonable aspiration, but in reality it has to be balanced with the associated cost in terms of monetary amount and the time it takes to obtain the curated and labeled data, validate the new model or to triage issues associated with the new retrained models.

As machine learning becomes deeply integrated into a company's core business workflow and online products, and as it continues to demonstrate its values in areas like customer experience and revenue generation, the need to maintain and improve the performance of those models becomes increasingly important to the business. One effective approach to satisfy this objective is through continuous model training, which is the primary focus of MLOps level 1 in Google Cloud's MLOps guide.[8]

To be effective at continuous model training, it is highly recommended to figure out a way to seamlessly integrate it into the standard machine learning development workflow, and develop a strategy to implement continuous model training that meets the company's specific needs and use cases. Regarding the strategy, there are three aspects to consider:[9]

- When: This is referring to under what conditions should the machine learning pipelines get triggered to retrain the model.

- What: This is referring to deciding on the right amount of data, which includes both existing and new data, to use to retrain the models.

---

[8] Google Cloud MLOps: Continuous delivery and automation pipelines in machine learning, `https://cloud.google.com/architecture/mlops-continuous-delivery-and-automation-pipelines-in-machine-learning`

[9] Framework for a successful Continuous Training Strategy, `https://towardsdatascience.com/framework-for-a-successful-continuous-training-strategy-8c83d17bb9dc`

- How: This is referring to a more comprehensive evaluation of whether to also adjust any hyperparameters or iterating on a new model version as a part of the model retraining exercise with the new data.

Among these three aspects, the one that model training infrastructure has more opinions on and can directly provide infrastructure to support for is the first one. The last two aspects are typically determined by the model owners, as they possess the best understanding of the specific machine learning use case, model architecture, and the current state of their model's evolution.

Regarding when best to trigger the machine learning pipelines to retrain models, the three common approaches are periodic retraining, model performance, and data change.

- Periodic retraining: This is a simple and straightforward time or cron-based approach. It is both easy to reason about and to implement. This approach works best when the frequency aligns with the data change pattern. If they are not, then the model retraining might be unnecessary and wasteful in terms of time and compute resources.

- Model performance: This approach relies on empirical evidence of model performance degradation. This approach is more suitable for machine learning use cases where obtaining ground truth quickly is feasible, such as the advertisement click-through rate prediction. However, it is less suitable for use cases where obtaining the ground truth takes a longer time, such as loan approval prediction.

- Data availability or change: This is a proactive approach of improving model performance by retraining the models when the new data is available or there is change in the data distribution. This approach is quite suitable for use cases where obtaining ground truth quickly is not feasible.

For some use cases, the best approach to trigger the machine learning pipelines to retrain models might be a combination of above approaches.

A modern orchestration tool can easily handle continuous model training based on either periodic retraining or data availability. However, for the other two approaches, model performance and data change, the model training infrastructure needs to provide mechanisms to track and monitor them. Then trigger the model training pipelines to retrain the models when those metrics fall below or exceed the configured thresholds.

144

Model Training at Scale

Model training becomes particularly challenging when one or more of the following requirements emerge in your machine learning projects:

- The dataset volume exceeds the storage capacity of a single computer. This is often the case with machine learning use cases in these areas: natural language processing, computer vision, and machine translation.

- The model is so complex that the number of parameters exceeds the memory capacity of a single computer. This is often the case for deep neural network models with many layers and parameters, such as GPT-3 or BERT models.

- The training time on massive datasets can be extremely time-consuming.

There is an obsession in the deep neural network community about training large and complex models with large datasets. One important reason behind this obsession is that research has shown that deep neural network models can achieve state-of-the-art results when they are trained on a large dataset.[10] This is especially true for machine translation or computer vision–related tasks. Figure 4-9[11] shows deep learning models outperform traditional machine learning models and the more data those neural network models are trained with, the better their performance is.

---

[10] Ilya Sutskever, Oriol Vinyals, Quoc V. Le, "Sequence to Sequence Learning with Neural Networks," 2014, https://arxiv.org/abs/1409.3215

[11] Adrew Ng, How Scale is Enabling Deep Learning, www.youtube.com/watch?v=LcfLo7YP8O4

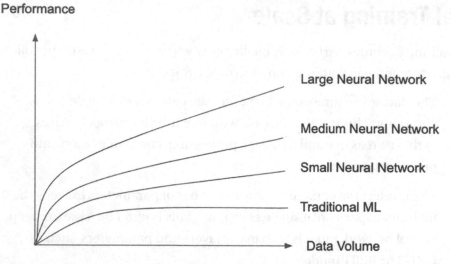

Performance

Large Neural Network

Medium Neural Network

Small Neural Network

Traditional ML

Data Volume

Figure 4-9. *Deep neural network model performance comparison*

For most enterprise and practical machine learning projects, the last requirement among the above three is the most common. Data scientists rely on the ability to iterate quickly during model development to produce high-performing models. To meet this challenge, the model training infrastructure should offer tooling and compute resources, enabling data scientists to leverage distributed model training.

Distributed Model Training

Distributed model training, mainly applicable for Deep Learning, combines distributed system principles with machine learning techniques to train models on a cluster of machines with GPUs by distributing the training workload across those machines and ensuring that the model training is fault-tolerant and scalable. The primary objectives of distributed model training are to speed up the training process, and to facilitate the training of models that are too large or complex to be trained on a single computer.

The two main approaches to distributed model training are data parallelism and model parallelism. The intricacies of these approaches are beyond the scope of this book, and here are the simplified explanations of these approaches.

In data parallelism, the training data is split into smaller batches and distributed across different machines, and each machine trains the model on its own batch of data, as depicted in Figure 4-10. The results are then combined to produce a single model. This approach works with most deep neural network model architecture, and that is why it is much more widely adopted.

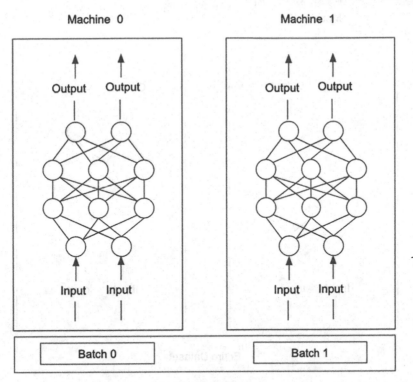

Machine 0   **Machine 1**

Entire Dataset

Figure 4-10. *Data parallelism*

In model parallelism, the large model is divided into smaller parts and distributed across different machines, and each machine trains its own part of the model using the same entire training data, as depicted in Figure 4-11. This approach works best when parts of the deep neural network model architecture can be independently computed in parallel.

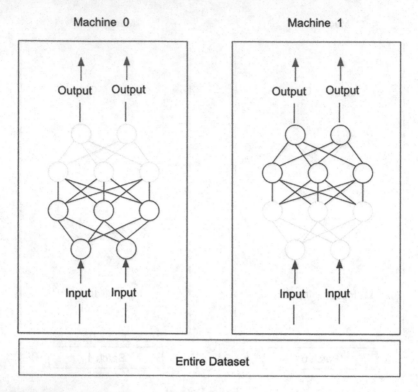

Figure 4-11. *Model parallelism*

Fortunately, most popular deep learning libraries like TensorFlow and Pytorch have built in support for distributed training.

From the model training infrastructure's perspective, it should provide the cluster management in terms of spinning up and tearing down the clusters in a self-service and efficient manner. This is where Kubernetes comes into play. Modern model training infrastructures now rely on Kubernetes to provision and orchestrate the model training workload, ensuring the availability of appropriate compute resources, including the number of nodes, CPUs, and GPUs. Additionally, Kubernetes offers a clean isolation for different model training jobs, enabling each one the freedom to use necessary libraries or specific library versions to meet their needs, such as exploring new features in one of the distributed learning libraries.

To help accelerate the model exploration and development steps, the clusters should be set up with commonly used deep learning libraries and toolings to help data scientists understand the model training progression as well as the cluster usage efficiency.

The LyftLearn architecture[12] from the Lyft Machine Learning platform team provides a comprehensive overview of their model training infrastructure. It outlines the essential components and their interactions, serving as a valuable blueprint for those building their down model training infrastructure. While the high-level view may appear straightforward, it is important not to underestimate the complexity and effort required to turn such architecture into a reality. A well-known saying goes, "the devil is in the details," and this certainly holds true in this case.

Model Registry

The output of the model training step is machine learning models and their associated metadata. Over time, their number will grow as more machine learning projects are undertaken and more data scientists join the team. As the model count increases, manually tracking and discovering them, and managing their lifecycle in a manual and ad hoc way becomes increasingly challenging, time consuming, and error prone. This is when adopting a model registry becomes a sensible choice. In highly regulated industries, such as finance services, healthcare, and insurance, model registry can help with model governance and validation to ensure proper model deployment process is followed with detailed auditing and tracking.

At the high level, the model registry is a crucial component of MLOps, providing a structured and controlled environment for publishing machine learning models and managing their lifecycle. Its role is analogous to that of Artifactory or Docker Hub in software development. Furthermore, it plays an important bridging role[13] between the machine learning development and operationalization phases of the machine learning development lifecycle, as depicted in Figure 4-12.

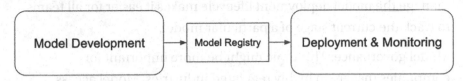

Figure 4-12. *Model registry as a bridge*

---

[12]Vinay Kakade, LyftLearn: ML Model Training Infrastructure built on Kubernetes, https://eng.lyft.com/lyftlearn-ml-model-training-infrastructure-built-on-kubernetes-aef8218842bb

[13]Simplifying MLOps with Model Registry, www.youtube.com/watch?v=WrieKPgXZyo&t=643s

Before diving into the technical aspects of a model registry, let's discuss the benefits it brings to the entire machine learning community inside companies. Notice, these benefits will pay big dividends when the scale of applying machine learning reaches a certain critical threshold, such as more than two machine learning teams or the number of models is greater than ten.

- Documentation and discovery: With storing the proper metadata and model artifacts in a centralized repository, data scientists, machine learning engineers, and others can easily discover the state and lifecycle of the various models developed by their teams or other teams in the company. If they are curious, they can easily find out how those models were trained, and by whom.

- Collaboration: Various teams involved in managing model lifecycle or debugging model-related issues can easily collaborate when they all have access to the same information about the various models in a centralized place.

- Deployment: Model registry acts as a bridge between machine learning experimentation during model development and operationalization, which might be the responsibility of the operation team and not data scientists. Only the models that were published to a model registry will be tested, validated, and deployed to production, and not some random models from a data scientist's laptop.

- Lifecycle management: Certain model registries offer predefined or configurable model deployment phases. Having the model registry manage the model deployment lifecycle makes it easier for all teams to track the current stage of a particular model.

- Model governance: This topic might be more important for companies that are in highly regulated industries. Model access controls, deployment permissions, auditing reports, and model lineage are important aspects of responsible AI governance best practices.

Note   Model registry, model store, model repository

As an MLOps industry, we are very good at coining terminologies for concepts and ideas, and sometimes this leads to confusion due to overlapping definitions and a lack of oversight. A quick Internet search reveals that these terms often appear closely related, and at times, even synonymous. While there might be minor differences, it is more important to focus on understanding the roles, benefits, and high-level technical details associated with the management of machine learning models, their metadata, and their lifecycle within the context of MLOps.

Much like the maturity level of experiment tracking solutions, both the open source and vendor-based model registry solutions have reached a significant level of maturity. We now have a wide range of choices available to choose from. Considering the current state of these model registry solutions and if your organization needs one, it is highly advisable to adopt one rather than building one internally, unless there are very specific requirements that none of the existing solutions can meet.

MLFlow has a component called "Model Registry." Once the models are registered, then they show up in the "Registered Models" table, as depicted in Figure 4-13.

The model version detail screen will display the model metadata and the associated experiment run, and Figure 4-13 is an example of what it looks like.

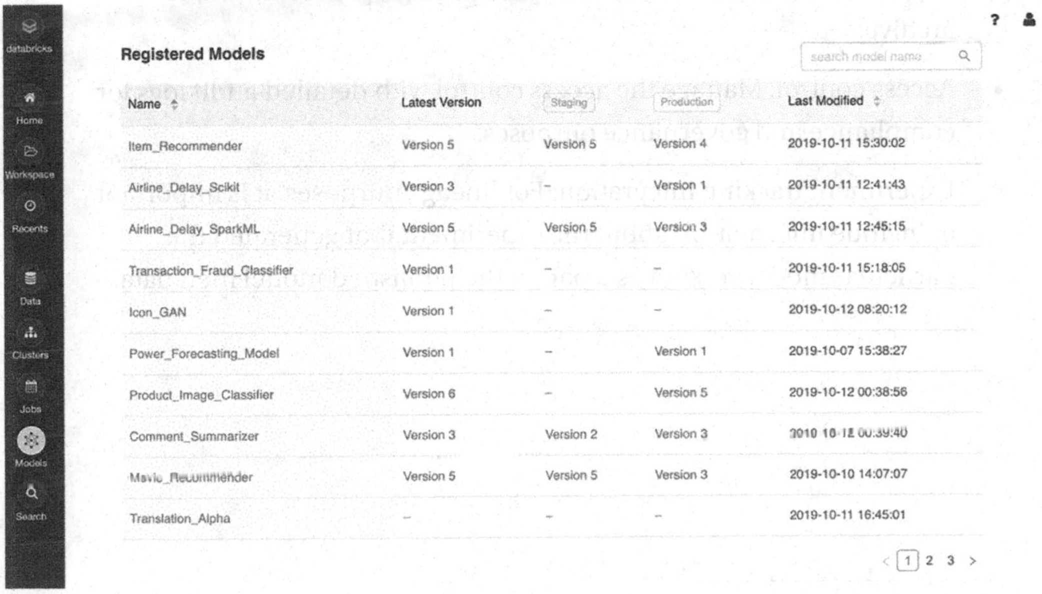

Registered Models				
Name ⇕	Latest Version	Staging	Production	Last Modified ⇕
Item_Recommender	Version 5	Version 5	Version 4	2019-10-11 15:30:02
Airline_Delay_Scikit	Version 3	–	Version 1	2019-10-11 12:41:43
Airline_Delay_SparkML	Version 5	Version 5	Version 3	2019-10-11 12:45:15
Transaction_Fraud_Classifier	Version 1	–	–	2019-10-11 15:18:05
Icon_GAN	Version 1	–	–	2019-10-12 08:20:12
Power_Forecasting_Model	Version 1	–	Version 1	2019-10-07 15:38:27
Product_Image_Classifier	Version 6	–	Version 5	2019-10-12 00:38:56
Comment_Summarizer	Version 3	Version 2	Version 3	2019-10-12 00:39:40
Movie_Recommender	Version 5	Version 5	Version 3	2019-10-10 14:07:07
Translation_Alpha	–	–	–	2019-10-11 16:45:01

< 1 2 3 >

Figure 4-13. MLFlow registered models

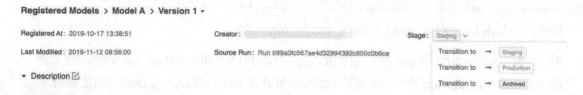

Figure 4-14. *Metadata and experiment of registered model*

This model registry–related blog[14] not only shares the best model registry tools and how they compare, but also outlines the evaluation criteria to help readers decide on the best one for them.

High-Level Architecture

While some of the internal components in the model registry architecture are quite similar to the ones in the experiment tracking architecture, there are few key pieces to pay attention to:

- Downstream service integration: Provide a scalable and easy means for downstream services to consume or download the machine learn model through API integration.

- State transition: Record and manage the state transition of the model deployment lifecycle from development, staging, production, to archive.

- Access control: Manage the access control with detailed audit logs for compliance and governance purposes.

- Experiment tracking integration: For lineage purposes, it is important to include information about the experiment that generated the particular model version as a part of the published model metadata.

---

[14] Best ML Model Registry Tools, https://neptune.ai/blog/ml-model-registry-best-tools

Figure 4-15 depicts a high-level architecture of model registry. Besides the standard service to handle incoming requests in either HTTP or gRPC protocol from various clients and interacting with various backend storages for structured metadata and models, internally, it might need sophisticated access control management logic and potentially an integration with internal LDAP system for authentication and roles. Additionally, it might need sophisticated logic for accurately tracking and transitioning models through either a predefined or configurable lifecycle.

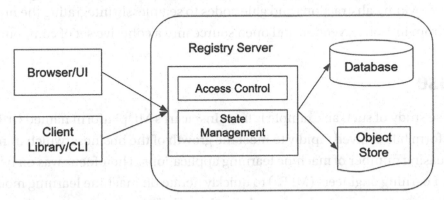

Figure 4-15. *Model registry high-level architecture*

The model registry is an important component of the model training infrastructure, acting as a bridge between the model development and operationalization phases of the machine learning development lifecycle.

Case Studies

The model training infrastructure plays a key role in enabling data scientists and machine learning engineers to speed up the model development and exploration phases while promoting best practices around collaboration, reproducibility, and automation. The necessary sophistication of a model training infrastructure depends on various factors, including the number of active machine learning use cases, the number of data scientists, the scale and complexity of the machine learning models, and the extent to which machine learning is integrated into product development within a company.

For startups or organizations that are at the beginning of their journey in applying machine learning, it is quite sensible to leverage the cloud provider solutions for most, if not all, of their model training infrastructure needs. Another option is to combine the best of cloud provider solutions, in-house solutions, and selective vendor solutions.

For medium-sized organizations with many machine learning use cases, a sizable number of data scientists, and machine learning is an integral part of product development, then the focus is more on extensibility, scalability, efficiency, and self-service by providing abstractions and glue codes to seamlessly integrating the numerous solutions from in-house, vendor, and open source into a cohesive set of components.

In-House

A good case study of such an example is from Instacart's ML platform named Griffin. Their platform has evolved rapidly to meet the growth of the business, which translates to an increasing number of machine learning applications. Their focus was on helping "Machine Learning engineers (MLEs) to quickly iterate on machine learning models, effortlessly manage product releases and closely track their production applications."[15] With this in mind, they set out to build their model training infrastructure with these major considerations:

- Scalability: The platform must have the capability to accommodate thousands of machine learning applications.

- Extensibility: Flexibility and extensibility are a must to support various machine learning tools and other needed peripheral systems in the machine learning development process.

- Generality: The platform should offer a unified workflow and a consistent user experience while abstracting away the broad integration with third-party solutions.

Figure 4-16 depicts a simplified version of the Griffin system architecture and mainly focuses on model training infrastructure–related components.

---

[15] Instacart's hyper-growth entails increasing machine learning applications and requires fast iterations of machine learning workflows, www.instacart.com/company/how-its-made/griffin-how-instacarts-ml-platform-tripled-ml-applications-in-a-year/

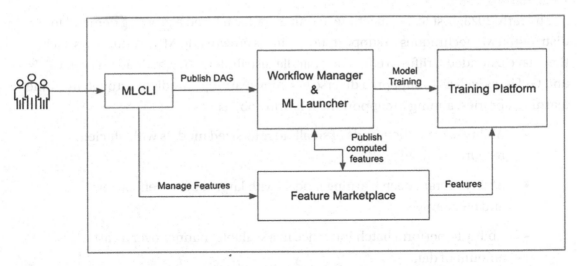

Figure 4-16. *An adaptation of Griffin System Architecture from their engineering blog*

The MLCLI is the main interface the MLEs use to interact with Griffin machine learning platform to develop their machine learning applications and manage model lifecycle. At the high level, it provides an easy way to get started by generating ML workflow code from base templates. MLEs then can customize the generated code to meet their needs, and finally they can submit their code in the form of pipelines to train their machine learning models.

In terms of machine learning pipeline scheduling and orchestration, Griffin utilizes Airflow as an orchestrator. The machine learning pipelines are defined in a declarative manner in YAML format. This abstraction hides the underlying intricacies of working with Airflow and enables MLEs to focus on defining their machine learning pipeline tasks.

To accommodate the diversity in machine learning frameworks, such as Tensorflow, Pytorch, scikit-learn, and more, Griffin provides a framework-agnostic training platform that standardizes package management, metadata management, and code management. To train their models, MLEs begin by selecting the training framework suitable for their use case and then provide the model training logic and hyperparameters. The underlying machinery of the model training platform handles the provisioning of the required compute resources and launches the model training jobs.

To support experiment and metrics tracking during the model training process, MLFlow is used to manage and store them as well as for managing models, serving as a model registry.

155

In early 2023, Instacart's machine learning platform faced a growing demand for distributed ML techniques to support the business growth and ML products. As such, they have extended Griffin to efficiently handle distributed ML workload on both CPUs and GPUs, as well as to support a diverse machine learning paradigm through machine learning libraries, aiming to support the following objectives[16]:

- Ability to train thousands of small- to mid-sized models with efficient resource utilization

- Ability to train deep learning models with large datasets effectively and efficiently

- Ability to perform batch inference in a scalable manner over a vast amount of data

Ray,[17] an open source unified compute framework that makes it easy to scale AI and Python workloads, was chosen as the foundational computation framework for their distributed machine learning training platform. Figure 4-17 depicted the high-level distributed model training architecture and what the flow looks like during the interactive and automated model development. The main aspect to highlight is that the same backend components that consist of the Griffin Workflow Control Plan and Ray Cluster running on Elastic Kubernetes (EKS) are used to support both flows.

---

[16] How Instacart uses distributed Machine Learning to efficiently train thousands of models in production, www.instacart.com/company/how-its-made/distributed-machine-learning-at-instacart/

[17] Effortlessly Scale Your Most Complex Workload, www.ray.io/

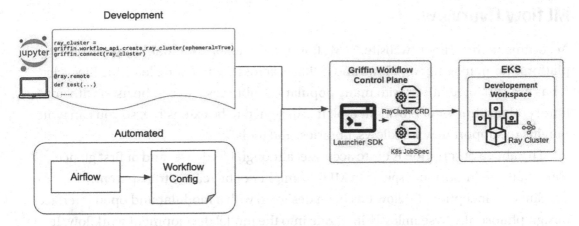

Figure 4-17. *An adaptation of how the Griffin workflow control plane and Ray Cluster are used in interactive and automated model training[17]*

Open Source

One of the most popular and widely adopted open source projects in the MLOps ecosystem is called MLflow, often described as an open source platform for managing the end-to-end machine learning lifecycle. It was created by the team at Databricks in mid-2018, and the 1.0 version was released one year later in mid-2019.[18] Over the last few years, MLflow has steadily matured in terms of capabilities, and its adoption has dramatically increased with over ten millions monthly downloads.[19] At the time of this writing, MLflow version 2.7 has a few LLM-related features, although they are in the experimental stage.

The following sections will provide an overview of MLflow, its main feature set, and high-level architecture.

---

[18] MLflow 1.0 release, https://mlflow.org/category/news/index.html

[19] 10 MLflow Features to 10 Million Downloads, www.linuxfoundation.org/blog/10-mlflow-features-to-10-million-downloads

MLflow Overview

According to the MLflow website,[20] "MLflow is a versatile, expandable, open-source platform for managing workflows and artifacts across the machine learning lifecycle. It has built-in integrations with many popular ML libraries, but can be used with any library, algorithm, or deployment tool. It is designed to be extensible, so you can write plugins to support new workflows, libraries, and tools."

The above description is quite accurate, although it is dense, and at first glance, some of those important aspects in MLflow might not be readily transparent.

Since its inception, MLflow has been designed with a modular and open interface design philosophy to seamlessly integrate into the model development workflow. It provides command-line interfaces, REST APIs, and a user interface to easily integrate with various steps in the model development phase, including experiment tracking, model packaging, model deployment, and more.

It is quite easy to get MLflow up and running on a local machine using just a few commands after downloading it from their website. For production usage across multiple teams, it will require a bit more investment to configure MLflow to use persistent storages, and some of these details will be covered in the following.

MLflow Feature Set

MLflow version 2.7 offers five production-ready components (see Figure 4-18) and two experimental components. Adopting MLflow is not an all-or-nothing decision; instead, you have complete freedom to pick and choose which components to start with.

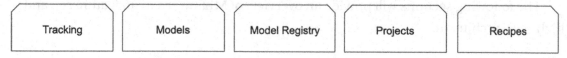

Figure 4-18.   *MLflow five production-ready components*

The two notable components are Tracking and Model Registry, which data scientists and MLEs most frequently use and interact with. The other three components are primarily focused on providing best practices to aid in packaging machine learning

---

[20] What is MLflow? https://mlflow.org/docs/latest/what-is-mlflow.html

projects, models, and defining consistent machine learning workflows. In the context model training infrastructure, the Tracking and Model Registry components are particularly relevant, and their capabilities are discussed below.

The two experimental components are designed to support the emerging needs in the large language model (LLM) era: AI Gateway and Prompt Engineering UI. The first one is intended for organizations to use as a centralized service to streamline interactions and the management of various LLM providers, such as OpenAI, Google, Cohere, and Hugging Face. Once the AI gateway service is up and running, we can leverage the second experiment component, called Prompt Engineering UI, to easily test the same prompt against multiple LLM providers and evaluate their responses.

MLflow Tracking

This component is often the first one that users explore when evaluating MLflow for adoption. Its primary focus is to fulfill the needs of experiment and metadata tracking during the model exploration and development phase. This enables data scientists to move away from using the traditional, manual, and error-prone methods, such as spreadsheets, or similar approaches.

The central concept in MLflow Tracking is called "*runs*," which can be grouped under experiments. Each run can be as simple as the execution of a piece of machine learning code or as complex as an end-to-end model training run with a set of hyperparameters. During these runs, we can use APIs to log various parameters, including code versions, metrics, output files, or any other useful artifacts.

Once data scientists have completed multiple runs to experiment with various hyperparameter combinations or different feature sets, they use the provided Track UI to visualize, search, and compare the runs that are in the same experiment. If necessary, they can also download the run results. Figure 4-19 shows an example of Tracking UI.

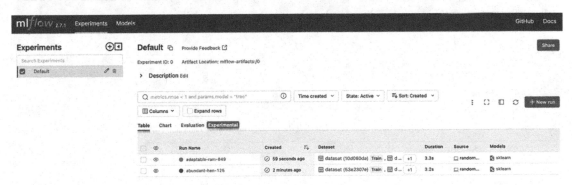

Figure 4-19. *MLflow Tracking UI in table format*

One of the most exciting features in the MLflow tracking component is automatic logging. This feature enables data scientists to log metrics, parameters, and models without the need to scatter log statements throughout the model training code. In other words, there is no need for explicit log statements. Out of the box, this auto-logging feature supports commonly used machine learning libraries, such as Scikit-learn, Keras, Pytorch, XGBoost, LightGBM, and more. Listing 4-1 demonstrates a short scikit-learn autolog example from MLflow website.

Listing 4-1. A short scikit-learn autolog example

```
import mlflow
from sklearn.model_selection import train_test_split
from sklearn.datasets import load_diabetes
from sklearn.ensemble import RandomForestRegressor

mlflow.autolog()

db = load_diabetes()
X_train, X_test, y_train, y_test = train_test_split(db.data, db.target)

# Create and train models.
rf = RandomForestRegressor(n_estimators=100, max_depth=6, max_features=3)
rf.fit(X_train, y_train)

# Use the model to make predictions on the test dataset.
predictions = rf.predict(X_test)
autolog_run = mlflow.last_active_run()
```

The Tracking UI in Figure 4-20 shows the comparison of two runs of the above code with different n_estimators and max_depth parameter combination.

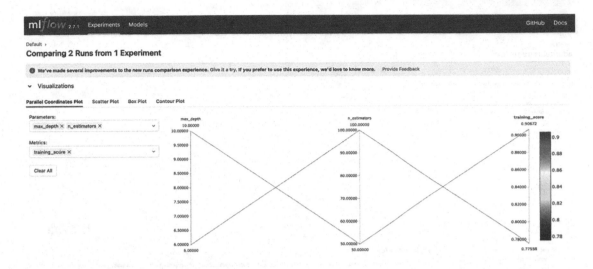

Figure 4-20. *MLflow Tracking UI compares training score of two runs with different parameters (n_estimators and max_depth)*

MLflow Model Registry

The MLflow Model Registry is designed to assist data scientists and MLEs to collaboratively manage the lifecycle of their models by centrally storing models along with their associated metadata. This metadata includes details such as the MLflow experiment and run that produced the model, the version, annotations, and state transitions. All interactions with this component are performed through the provided APIs and UI.

The first step in interacting with this component is to register the models using one of the aforementioned methods. A registered model is assigned a unique name and includes the various pieces of metadata mentioned above. The MLflow Runs detail page provides a "Registered Model" button for easy model registration via the UI, as depicted in Figure 4-21.

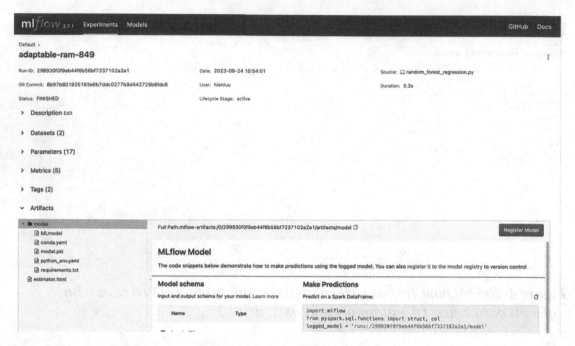

Figure 4-21. *Register a model using MLFlow Run Details UI*

Data scientists often train multiple versions of the same model, each with a slightly different set of hyperparameter combinations. A registered model can have multiple versions in the MLflow Model Registry, as illustrated in Figure 4-22, which depicts an example of this scenario.

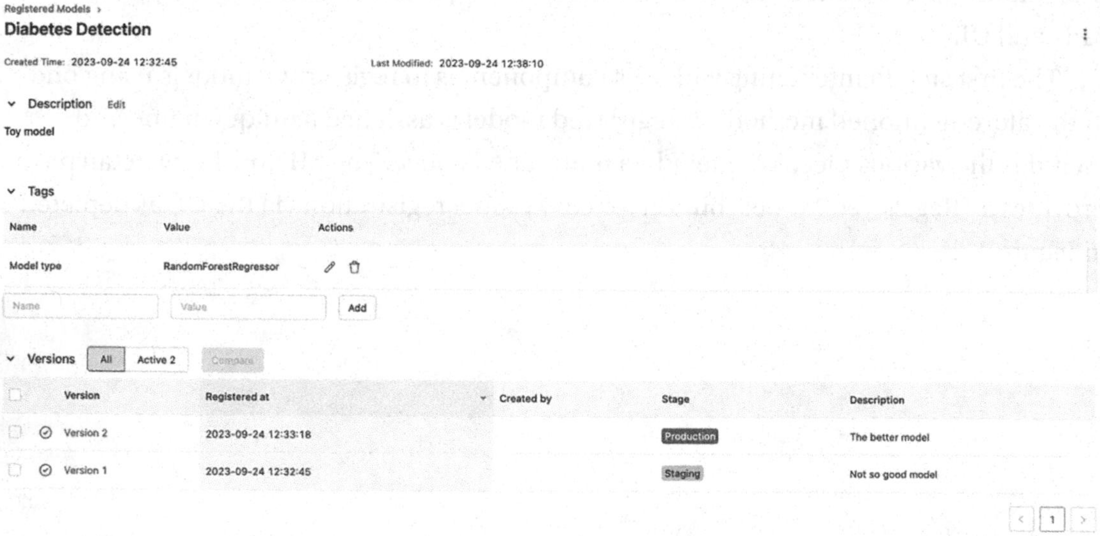

Figure 4-22. *A registered model with two versions in MLFlow Registered Models UI*

In terms of model stages, MLflow provides predefined stages for common use cases, such as Staging, Production, and Archived. At any given time, each distinct model version can be assigned one stage, meaning that no two model versions of the same model can be in the same stage. The ability to define custom stages is one of the current open feature requests on MLflow's GitHub.[21]

MLflow High-Level Architecture

At the high level, MLflow system logical architecture, as depicted below in Figure 4-23, is a standard three-tiered architecture, which has three parts: MLflow client, tracking server, and backend stores.

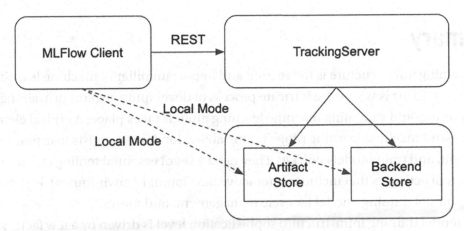

***Figure 4-23.** The three main components in the MLflow system logical architecture*

The MLFlow client uses the MLflow library to interact with the TrackingServer via REST HTTP protocol. An MLflow client could be a model training script or an application that interacts with TrackingServer to query, update, or insert run information or model registry–related operations.

The TrackingServer acts as a REST service that performs operations on Backend Store and potentially Artifact Store, depending on the type of configuration. When using MLflow in local mode, the MLflow client will interact directly with local Artifact Store and Backend Store.

---

[21] Allow users to define custom stages for the model registry, `https://github.com/mlflow/mlflow/issues/3686`

MLflow records the metadata, params, metrics, and other details about the experiments and runs in the Backend Store, which supports two types of storage: file store or database-backed store. File store is mainly for a single person usage or during the exploration phase. For production usage, consider setting up a property instance of MySQL or Postgres to persist the data. Artifact Store is used for storing artifact outputs, such as models, images, and other unstructured data. For production usage, consider using one of those distributed storage services like AWS S3, Azure Blob Storage, or Google Cloud Storage.

The MLflow Tracking component provides many different options to record the runs and artifacts to meet the needs of a single user to large organizations. Those options are described in details in their documentation.[22]

Summary

Model training infrastructure is the second and important pillar of machine learning infrastructure. This is where the intricate process of developing, exploring, training, experimenting, and validating machine learning models takes place. A critical element of a successful machine learning project is the ability for data scientists to explore, experiment, and train models quickly. They need a set of essential toolings to support a diverse set of activities that includes interactive development environment, experiment tracking, model training, model lifecycle management, and more.

The model training infrastructure sophistication level is driven by a few factors: the number of machine learning use cases, the number of data scientists, the complexity of the model, the associated data volume to train models with, and the aspired MLOps maturity level. At the minimum though, it should provide the following components:

- Interactive development environment, such as Jupyter notebooks

- Experiment and metadata tracking

- Model training

- Model pipeline orchestration

- Model registry

---

[22] How runs and artifacts are recorded, `https://mlflow.org/docs/latest/tracking.html#how-runs-and-artifacts-are-recorded`

In recent years, numerous innovations have emerged in the area of MLOps toolings, coming from both the open source community and vendors. At the moment, organizations have many available options to choose from of the aforementioned components. While the build vs. buy discussion has become simpler than before, however the decision-making process for a specific solution remains a complex task. For organizations that are interested in exploring open source solutions, MLFlow should be considered as an option for their experiment tracking and model registry needs.

Irrespective of their size, organizations with a small number of machine learning use cases should consider starting by leveraging vendor-managed solutions. When the number of machine learning use cases exceeds ten and machine learning becomes an integral part of their product development, it may then make sense to invest in building an abstraction layer to hide the collection solutions from open source, vendors, and in-house development. This approach ensures a consistent experience for data scientists, future-proofs the underlying infrastructure, and facilitates future efficiency improvements.

Once the aforementioned set of infrastructure components is available and operational, achieving MLOps level 2[23] becomes significantly more attainable. This maturity level is the target level for organizations that are committed to harness the full power of machine learning. At this maturity level, organizations can effectively handle rapid changes in the data or business environment by establishing ML pipelines that automate the retraining and deployment of new models with minimum human intervention.

---

[23] MLOps: Continuous delivery and automation pipelines in machine learning, https://cloud.google.com/architecture/mlops-continuous-delivery-and-automation-pipelines-in-machine-learning

CHAPTER 5

Model Serving Infrastructure

In the machine learning (ML) community, there is a common saying that the ROI of an ML project starts when the model is in production. This phrase reminds us that the true value of a ML project is realized when the trained model is deployed and actively used in production. The model serving infrastructure plays a crucial role in operationalizing ML models in production and integrating the ML projects into the operations of an organization, such as predicting customer churn, detecting fraudulent activities, personalizing customer experience, improving the quality of products and services, and more.

The model serving infrastructure is vast and complex compared to the other infrastructures within the ML platform. It requires a significant amount of software engineering, and there are numerous challenges, especially when it comes to performing real-time inference at scale and supporting a sizable number of ML models that are constantly deployed and experimented by multiple teams within an organization.

Note Model inference vs. model prediction

Model inference and model prediction are two closely related concepts and often used interchangeably; however, their meaning is slightly different.

The model inference concept is about the process of using a trained ML model to make or generate predictions or decisions on new data the model has not seen before.

© Hien Luu, Max Pumperla and Zhe Zhang 2024
H. Luu et al., *MLOps with Ray*, https://doi.org/10.1007/979-8-8688-0376-5_5

The model prediction concept refers to the output or decision generated by a model, which could be a probability score, a classification label, and more. The generated prediction is based on the learned relationships between the input and output data.

In short, the model inference is about the broader process of using the model to make predictions based on new data, and model prediction refers to the output generated by the model.

The model serving infrastructure provides the necessary infrastructure to perform model inference using the trained and validated models on new, unseen data in order to make predictions, whether those are classifications or regressions. This is the pivotal point in the machine learning development process where the trained models are deployed to production to perform model inference.

There are two standard model inference paradigms: offline and online, each catering to different set model inferences needs. Offline inference is used to generate the model's predictions in batch mode using a batch of data. Online inference is commonly employed to generate the model's predictions online or in real-time, especially in use cases where ML models are used to provide personalized and relevant experiences to users, such as recommending movies, detecting fraudulent activity, estimating the food arrival times, and many more. In these cases, model inference is an integral part of handling user interactions, and as a result, it must be performed with low latency and high reliability.

In this chapter, we will first discuss the high-level architecture of the model serving infrastructure to support the two inference paradigms mentioned above. Then, we will explore the various design decisions for building a scalable, reliable, and efficient system to support large-scale online inference and supporting a large number of diverse ML models. For the last part of the chapter, we will highlight a few notable open source projects that are suitable for building model serving infrastructure on top of, and we will discuss a few in-house solutions from large companies to learn from.

Overview

The model serving infrastructure is the key enabler for bringing ML models to production quickly and easily. The process of getting the model to production, and ultimately to model inference, consists of many steps, including model deployment, model preparation, model serving, model monitoring, and model decommissioning. The following is a high-level description of each of these steps:

- Model deployment

  - This step is about properly packaging the trained model with the necessary runtime configurations and metadata, transferring it to the model registry and triggering the CD pipeline for the actual deployment.

- Model preparation

  - This step is about loading the model deployment package from the model registry into the model serving service.

- Model serving

  - This step is about serving the model inference requests from clients by preparing the necessary features and providing them to the appropriate ML model to perform predictions.

- Model monitoring

  - This step is about ensuring proper logging of both ML model performance metrics and model serving operational metrics for troubleshooting and performance optimization purposes.

- Model decommissioning

  - This step is about stopping model serving request for the decommissioned model and archiving the model deployment package in the model registry.

It is important to note that the descriptions above are meant to capture the general details; the specific aspects and nuances involved will vary depending on the type of model, serving infrastructure, and the organization's policies and procedures.

For small startups or organizations with only a handful of simple ML use cases, it may make more sense to leverage the cloud provider solutions, for most, if not all, of their model serving infrastructure needs.

For other organizations, here is a short list of considerations when deciding to incrementally invest in building out their model serving infrastructure using a combination of open source, home-grown, or vendor solutions:

- Scaling: When the number of ML use cases is growing and reaching double digits

- Real-time inference: When most of the ML use cases need real-time inference

- Widespread: When ML is becoming a crucial part of their product offerings

- Business value: When ML is making a significant impact on their business

Once the model serving infrastructure is in place, it will provide compounding benefits over time.

These are the common benefits that model serving infrastructure provides:

- Reduced time to production

  - The model serving infrastructure provides the commonly needed infrastructure and tools for model deployment and model inference. This can help teams to deploy ML model to production quickly and easily.

- Ease of integration

  - The model serving infrastructure typically provides APIs and client libraries that make it easy for product engineer teams or others to integrate the model inference requests into their applications, microservices, or systems.

- Scalability

  - A properly designed stateless model serving infrastructure running on Kubernetes can easily horizontally scale to handle the large inference workload. This allows teams to deploy ML models to production with very little concerns about the capacity constraints.

- Reliability and availability

  - The model serving infrastructure is typically designed to provide high reliability and availability through mechanism like load balancing and redundancy to support critical applications that need ML model predictions to make decisions or provide services.

The following section will delve in the details of the various components in the model serving infrastructure, the scaling challenges, the various engineering design decisions, and best practices. The discussions will be particularly relevant to use cases that demand scalability, reliability, and low latency in their model inference.

High-Level Architecture

An effective model serving infrastructure must provide a set of components and best practices to guarantee the realization of the benefits mentioned above.

Like most system architecture, there is no single perfect architecture to satisfy all the various requirements. The architecture that works well at one company might not necessarily be suitable for another company. This is because the architecture of a system is heavily influenced by the various considerations and trade-offs, which are derived from the specific system requirements and the environment it needs to operate in.

When designing the architecture of a typical model serving infrastructure, several common considerations come into play, including flexibility, model rollout approach, efficiency, latency, scalability, and reliability.

- Flexibility

  - This is referring to the support of pre-processing and post-processing of model predictions inside the model serving infrastructure. This support means running the user provided code inside the model serving infrastructure, which can potentially open a pandora box of challenges, including performance issues, operational complexity, and dependency conflict.

- Model rollout approach

  - Production deployment of a new model or the second and subsequent version of an existing model typically requires guardrails to minimize unexpected or surprising problems. The two common model rollout approaches are shadow and canary deployment.

- Efficiency

  - The efficiency in terms of compute resources, memory, and network bandwidth will become important when supporting a large number of ML use cases and models. These efficiencies will have a direct impact on the overall operational costs. The efficiency consideration will influence the decision around exposing one end-point per model and a single generic end-point for all one or more models.

- Latency

  - The common factors that have large influence on the model inference latency are model size, model complexity, and the number of needed features to generate predictions. Use cases like ads targeting or ecommerce personalized recommendations require low model inference latency. Meeting the low latency requirements under 100ms will require a combination of techniques.

- Scalability and reliability

  - These considerations are key to a successful model serving infrastructure as they have a direct impact on the performance and availability of ML models in production. It is critical to ensure the model serving infrastructure can seamlessly scale to accommodate the increased workload easily, while maintaining high availability, ensuring reliable model predictions to deliver business value.

Note   Shadow vs. Canary model deployment

Shadow model deployment is commonly used to evaluate the new model's performance and behavior in production without impacting the existing production model. This is usually done by having the shadow models generate predictions, but don't expose those predictions to users.

Canary model deployment is used to gain insights in how the new model performs on a smaller scale before exposing its predictions to all users. This approach minimizes the blast radius when something unexpected goes wrong with the new model.

Both shadow and canary model deployment are valuable techniques for deploying ML models to production with minimal risks. Choosing the most effective techniques depends on the model being deployed and the team's specific needs.

These considerations can be thought of as a spectrum, much like adjusting the volume on the speaker. We have the flexibility to choose the level or setting that best aligns with our specific needs and requirements. It is worth noting that as we move closer to the right or maximum level on this spectrum, the architecture will tend to be more complex, requiring additional resources and potentially increasing implementation intricacies and time.

From the engineering's perspective, a typical model serving infrastructure consists of only a few core components, as depicted in Figure 5-1. However, it needs to interact with many other components and infrastructures in order to carry out the end-to-end process of responding to model inference requests. Furthermore, when operating at scale and supporting many ML models and their variants, operating and managing them is no small task.

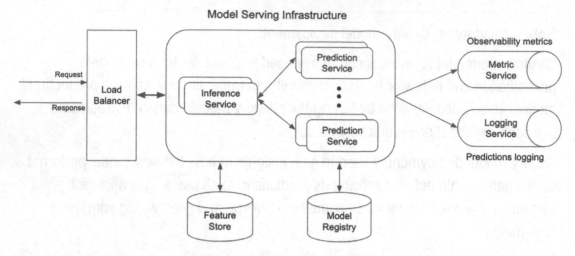

Figure 5-1. *Model serving infrastructure high-level architecture*

Before delving into the details of the core components of the model serving infrastructure, let's discuss the supporting roles of the surrounding components and infrastructures in Figure 5-1.

Feature Store

For organizations that have both offline and online ML use cases, it is common to store features in two places: the offline feature store and the online feature store. In the context model serving infrastructure, primarily designed for online ML use cases, it retrieves features from the online feature store during the inference process. This means the online feature store must provide low-latency access with sub-second response times. This requirement is one of the reasons why companies opt for key-value or NoSQL databases, such as Apache Cassandra, Redis, and Amazon DynamoDB, for storing and serving features.

To ensure feature freshness, the daily or hourly generated features are uploaded to the online feature store in an efficient and reliable manner.

Model Registry

In the context of model serving infrastructure, the model registry plays an important role of providing the ML model artifact and the model configuration for models that are ready to be deployed to production, as depicted in Figure 5-2.

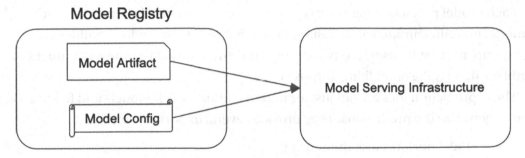

Figure 5-2. *Model registry supplying model artifact and model config*

Model artifact represents the trained ML model in binary format. The associated model configuration contains essential metadata and deployment configuration. The standard model metadata includes a unique model identifier, the model input features and their types, model type, and more. Likewise, the standard deployment configuration specifies the model deployment options, including aspects such as rollout strategy, traffic percentage, logging configurations, and more.

As part of the model deployment process, the model serving infrastructure receives notifications about models that are ready for use in production.

Metric Service

Effective management and operation of the model service infrastructure at scale requires insights into the various operational metrics. These metrics include aspects like latency, CPU and memory usage, network bandwidth, request rate, error rate, and more.

Similar to other complex microservices, the model serving infrastructure relies on comprehensive support from the metrics service. This support includes metrics aggregation, metric storage, metric visualization, and more.

Access to insights derived from these important operational metrics accelerate issue troubleshooting and ensure that the model service infrastructure remains scalable and reliable, meeting the demands of business needs for model inference.

Logging Service

Logging model predictions is a very critical part of the model inference process and shouldn't be an afterthought. It serves multiple purposes in the area of ML and MLOps, extending beyond mere auditing.

Each model prediction log entry typically contains essential information, such as model identifier, input feature values, prediction result, prediction confidence, timestamp, request id, user id, environment, and any additional context that might be useful for debugging or auditing purposes.

These prediction logs are not just for understanding which models and features were used to generate the predictions. They provide several important benefits:

- Model performance monitoring

  - Monitoring the model performance in real-time is a key to quickly identify model performance degradation. Prediction logging allows us to compare the predictions with the actual outcomes or detect the shift in model prediction distribution.

- Model evaluation

  - The model evaluation and performance assessment need the various model metrics, such as accuracy, precision, recall, F1 score, and more. The data to calculate these metrics from the information in the prediction log statements.

- Explainability

  - For some ML use cases, understanding why a model made a certain prediction is critical. The logging of prediction, along with additional context, will help with explainability.

- Regulatory compliance

  - In regulated industries, such as finance, healthcare, and more, for regulatory purposes, keeping model predictions may be required for compliance with data privacy and industry-specific regulations.

- Feature and model improvement

  - Analyzing the features and model performance in a production environment can help guide data scientists to make informed decisions about model retraining and optimizing the model performance of future model versions.

From the engineering standpoint, prediction logs are typically sent to the logging service provided by the infrastructure organization. A typical friction between the ML community and MLOps infrastructure team is about agreeing on an optimal balance between the amount of log entries to retain taking into consideration that costs are directly related to the log volume.

Once the log entries are persisted on either a data warehouse or a data lake, then the various metrics are calculated and aggregated to make it easier for the ML community to consume and perform analysis using either dashboarding solutions or plain SQL.

Inference Service

The inference service is responsible for serving the inference requests from one or multiple clients in a scalable, reliable, and efficient manner and with low latency. It provides this service to its clients through a contract called service endpoint, which defines the structure and format of the incoming requests and returned responses. More details about service endpoints are discussed in the following.

This service is typically architected using the microservice architectural pattern and deployed on a scalable container orchestration platform called Kubernetes. Depending on the requirements, such as rate of the request, latency, model complexity, feature volume, and more, the design of this service can be as simple as a single REST server or as complex as hundreds of nodes running on Kubernetes infrastructure.

Modern inference service typically provides a common set of capabilities, including service endpoint, inference request batching, feature fetching, model prediction, prediction logging, and more. Some of the important capabilities are discussed in the following.

Service Endpoint

One critical design decision for an inference service involves determining the service endpoint, also known as serving API. This endpoint establishes the contract between the clients and the service. The microservice domain offers numerous best practices regarding the design of a service endpoint. We should consider adopting these practices when designing the inference service endpoint, whenever applicable.

One consideration in designing service endpoints involves its generality. This factor is less important when dealing with fewer than a handful of models, but becomes crucial when supporting a growing number, ranging from 20 to hundreds of models. A recommended future-proof best practice is to design a single generic service endpoint capable of handling inference for all models. This approach not only simplifies the integration effort for the service clients, but also speeds up the end-to-end model deployment process. However, it comes with a trade-off – the inference request and response payload must be general enough to accommodate the diverse needs of various models, including the number of features, their types, their structure, and whether they are required or optional.

Another design consideration is about the service endpoint communication protocol. For low volume inference request rate (under thousands of requests per second), this consideration is less important, but becomes crucial when request rate is high (hundreds of thousands to millions of requests per second) and latency requirement is under 500 milliseconds.

Inference Request Batching

In the simple world, a client sends an inference request, and the server receives the request, performs inference, and sends back the inference response with the predictions. Everyone is happy.

In the real world of complex ML use cases for eCommerce recommendation, ride sharing, food delivery, and more, these platforms have millions of users using their platform daily. This translates to billions of inferences being performed every hour. One approach to optimizing inference latency and throughput is through request batching.

Two common strategies for request batching are client-side batching and server-side batching. In the client-side batching strategy, each client batches multiple inference requests and sends them over to the inference server in a single network request instead of one by one. This is extremely useful in situations where a client frequently sends many inference requests because it needs tens to hundreds of predictions to handle a single user interaction. In the server-side batch strategy, the server performs the batching of multiple inference requests without client's knowledge. This implies the request batching of multiple requests across multiple clients. This is particularly useful in situations with many clients, where each one sends relatively few requests each.

In the context of designing the inference request payload to support client-side batching, it must provide a flexible way for clients to express inference requests with or without batching.

Model Loading and Unloading

Similar to software, it is quite common for ML models to have multiple versions. The frequency with which new models are created can vary. There are multiple factors for creating multiple versions of a model, including model improvement after gaining additional insights about the problem, model improvement to adapt to the changing data or business rules, model improvement to meet regulatory compliance changes, and more. With the increasing number of models the inference service needs to support, it is very likely there will be multiple model deployments per day.

At the high level, model loading is the process of loading new models into the service in order to perform predictions. It is a critical part of the model deployment process. There are several common approaches to model loading: batch loading, on-demand loading, and hybrid loading:

- Batch loading

  - This approach loads all the assigned models all at once at the startup of the service. This implies that whenever there is a model deployment, the service will need to be redeployed or restart. Among the approaches listed in this section, this is the simplest one. This approach is suitable when the number of models is relatively small, and they don't change frequently.

- On-demand loading

  - This approach dynamically loads the new or updated models. This requires some sort of a polling technique to know if new or updated models are available to load. This approach is suitable when the number of models is large or the models change frequently.

- Hybrid loading

  - This approach combines the simplicity of the first approach and sophistication of the second approach. This approach is suitable when there is mix of frequently changed and infrequently changed models.

It might seem like model loading is an auxiliary capability; however, when it is properly designed, it can greatly help in bringing in models to production in a seamless manner.

Model unloading is typically performed when a particular model has reached the end of its lifecycle and it needs to be evicted. Depending on the adopted model loading approach, the model uploading is evicted accordingly.

Feature Fetching

Feature fetching is one of the important steps during the model inference process, and if not properly designed, it can significantly increase model inference latency.

During model predictions, the trained ML model takes the input features, applies learned patterns, and generates predictions. In the context of online inference, there are three options for handling feature fetching:

1. The client sends all the needed input features to the inference service.

2. The inference service fetches all the needed input features from one or more feature stores.

3. The client provides a subset of the needed input features, and the inference service fetches the remaining ones.

The third option is a hybrid approach that can be quite useful in some cases, such as when a small number of needed input features are calculated at real time based on user behavior. Additionally, feature fetching on the inference service side opens up the opportunity to cache commonly needed features across related ML use cases.

Model Prediction

Once the input features and the models are available, the next step in the inference process is to generate predictions. This entails feeding the features into the loaded ML models for it to produce predictions.

As a good practice, the logic for generating predictions is typically abstracted and implemented as a reusable library. This library can be effectively used in multiple scenarios, such online prediction, batch prediction, and others.

A design consideration is whether the actual prediction generation occurs locally within the inference service or remotely in a separate service, often referred to as the prediction service. If local execution is preferred, a straightforward in-process call to the prediction library is sufficient. Conversely, if remote execution is chosen, the inference service will make a remote call to the prediction service, which will be discussed below.

Prediction Logging

Prediction logging is about recording and storing the information about the predictions made by ML model during the model inference process. It includes information such as the input features and their values, prediction value, confidence score, model id, model version, feature fetching latency, prediction latency, timestamp, and much more. These pieces of information are extremely valuable to understand and monitor the model performance, to debug and troubleshoot expected behaviors, to understand the factors influencing predictions, to meet regulatory and compliance requirements, to create a feedback loop for improving the training data, and more.

For those reasons listed above, it makes sense that data scientists typically want to log all the predictions at all times. This can be easily achieved when the prediction log entry volume is low. When the prediction volume is high in billions of prediction log entries per day, then this can overwhelm the logging infrastructure and the cost will become a factor.

Supporting the option to specify the logging fraction as part of the model deployment metadata is good practice for the model inference service. This option empowers all stakeholders involved in model serving to easily adjust the logging volume, striking the balance between the need for prediction logs throughout the model lifecycle and preventing the logging budget from spiraling out of control.

Prediction Service

The prediction service is responsible for performing model predictions with the provided set of input features and returning the predictions. This service is typically packaged as containers and deployed with the necessary libraries, including the correct versions to support the ML models. It then makes sure they are successfully loaded into memory so they are available at prediction time.

There are a few important considerations when designing a prediction service to support a diverse set of ML use cases in an organization, including machine learning framework, cost effectiveness, prediction post processing, and more.

Machine Learning Framework

Tree-based models and deep learning models are two types of ML algorithms commonly used in a wide range of applications, including fraud detection, email sentiment analysis, and personalized recommendations. Popular ML libraries for training these model types include

181

- Tree-based model: Scikit-learn, XGBoost, and LightGBM

- Deep learning model: TensorFlow and Pytorch

The design complexity of the prediction service is influenced by the level of flexibility an organization would like to provide to their ML community. On one hand, the ML community has complete freedom in choosing which ML library to use to train ML models. On the opposite end, the ML community is limited to a predefined set of ML libraries. Each approach has its own pros and cons. From an engineering standpoint, the second option tends to be much simpler to maintain and support.

Cost-Effectiveness

Cost-effectiveness becomes important when supporting hundreds or thousands of ML models. One contributing factor to this is compute utilization efficiency, which is about ensuring the allocated CPUs or GPUs are used most of the time.

Figure 5-3 depicts two approaches to model allocation on prediction service. When the number of models is small, the "one model per prediction service container" approach offers several advantages. First, it provides complete isolation, ensuring misbehaving models don't impact others. Second, it simplifies support for multiple different ML frameworks and versions. For example, one model is using the Pytorch 1.x library version, and another one is using the Pytorch 2.x library version. One disadvantage of this approach is that the compute resource used for making predictions might be not efficiently utilized due to low volume prediction requests, leading to idle CPU.

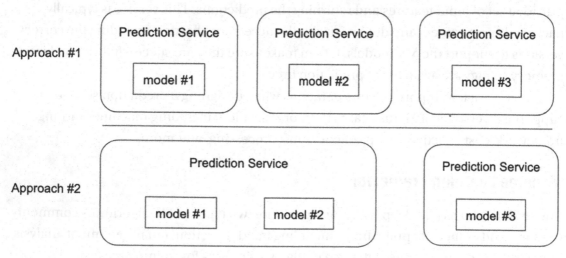

Figure 5-3. *Model allocation approaches*

The approach of hosting more than one model in a single prediction service container will help increase compute utilization. However, it comes with the trade-off about the potential noisy neighbor issues. This occurs when one model is hogging the CPU, perhaps due to high prediction request volume or model complexity, while other models are starved of CPU resources. In addition, managing library dependency and their version conflicts will be a challenge if those models depend on a variety of libraries with specific versions.

The hybrid option is commonly adopted in organizations where there are hundreds or thousands of models. This approach balances the advantages of the two aforementioned approaches while mitigating their respective drawbacks. However, implementing and maintaining this hybrid approach will require a larger team and increased effort. One organization that has thousands of models is Salesforce;[1] due to the strict requirement about data privacy, they train one model for each of their thousands of customers.

Prediction Pre-processing and Post-processing

Prediction pre-processing involves applying transformations to input features to ensure they are in a suitable format or value for the ML model to make more accurate predictions. This pre-processing step is usually aimed at improving the accuracy, reliability, and generalizability of the model's prediction. Here are ML use cases that often need prediction pre-processing:

- Image recognition: The provided image data might need to be resized, normalized, making it easier for the model to identify or classify objects accurately.

- Natural language processing: The raw text often will go through a few essential pre-processing steps, such as tokenization, stemming, and more, to ensure a consistent representation of the text. This will help improve the model's ability to make accurate predictions.

[1] Manoj Agarwal, Serving ML Models at a High Scale with Low Latency, 2021, www.youtube.com/watch?v=J36xHc05z-M

- Fraud detection: The combination of numerical and categorical data often needs transformation like imputing missing value or encoding categorical variables into numerical values. By transforming the input feature values into a standardized format, the model can be more effective at identifying patterns and anomalies that might indicate fraudulent activities or high-risk profiles.

Prediction post-processing is typically used to further refine, interpret, or transform the predictions to be useful for the intended application. Here are a few scenarios where often need prediction post-processing:

- Thresholding: In typical classification tasks, the predictions often represent confidence scores. One common post-processing step is about setting a threshold to determine the final classification label.

- Ensembling: In an ensemble learning technique, predictions from multiple models are combined to improve the overall accuracy. Prediction post-processing involves techniques like voting, weighted average to combine the individual predictions into a single output.

Supporting prediction pre-processing or post-processing or both in a generic manner requires the execution of custom logic provided by ML model owners within the inference service. From a software engineering standpoint, this flexibility introduces complexity and maintainability challenges, potentially increasing the difficulty in diagnosing erroneous prediction or identifying the latency-related issues.

Prediction Step Design Choice

In the inference process, a key step is the prediction step, which is responsible for generating the actual predictions based on the input data. This step can be computationally intensive, especially when dealing with complex ML models. In the design of a model serving infrastructure architecture, a key decision that must be made is whether to embed the model prediction step within the inference service or have it reside in a separate service. This decision should be based on a careful consideration of the specific requirements and constraints of the system, as well as the trade-offs involved in each approach. Some factors to consider include the complexity of the prediction step, the scalability and performance requirements of the system, and the need for flexibility and modularity. Let's discuss the pros and cons of each approach.

Embedded Prediction Step

In its most simplicity, Figure 5-4 depicts what it looks like in this design choice. In software design, a common and recommended wisdom we all follow most of the time is to keep it simple or start with simple. Let's unpack the advantages and disadvantages of this approach.

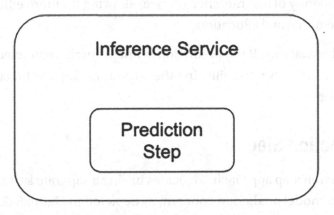

Figure 5-4. *Prediction step resides within the inference service*

The advantages are

- Simplicity and efficiency: It is quite easy to understand this approach simplifies the architecture since the overall number of components is reduced. When the prediction step lives inside the same inference process, this can lead to improved performance and reduced latency from minimal interaction overhead.

- Control: In general, it is much easier to control a step that lives within the same process than the remote one. In addition to a more efficient data transfer to prediction step, and a fine-grained control over the prediction process, the inference process has an easier time to handle error conditions and has more options to optimize performance.

- Deployment simplicity: Managing the deployment of a single service is definitely simpler than two services. Less artifacts, coordination, and potential for compatibility issues we need to worry about. This can be particularly beneficial for small teams with limited deployment resources.

185

The disadvantages are

- Scalability limitation: The model prediction step tends to be computationally expensive or requires large amounts of memory due to model size. In such cases, a separate service can be scaled independently of the inference service, allowing for more efficient resource usage and allocation.

- Limited reusability: If the prediction is not properly abstracted, this can restrict its reusability in other scenarios, such as batch prediction.

Remote Prediction Step

The remote prediction step approach advocates having a separate service be fully responsible for the model prediction concern, as depicted in Figure 5-5. In distributed system design, separation of concern is a design principle that involves dividing the system into different components that handle specific responsibilities. Designing the prediction step as a service follows the above design principle. At the same time, let's not forget the "no free lunch"[2] concept. Let's unpack the advantages and disadvantages of this approach.

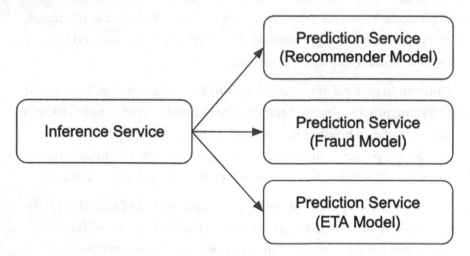

Figure 5-5. *Prediction step resides in the prediction service*

---

[2] David H. Wolpert and William G. Macready, "No free lunch theorems for optimization," 1997, https://ieeexplore.ieee.org/document/585893

The advantages are

- Scalability flexibility: A separate prediction service is particularly beneficial for computationally expensive models or scenarios of fluctuating or bursty prediction volume because we can scale this service independently.

- Improved modularity: This advantage comes directly from the separation of concern principle. Two important areas that can greatly benefit from the modularity are independent development and easier to test.

The disadvantages are

- Increased complexity: It is a generally acceptable argument that the over complexity of the system increases with an additional component. Introducing a separate prediction service increases the overall complexity of the model serving architecture. Specifically, this may lead to additional overhead in terms of deployment, maintenance, and communication between services.

- Performance overhead: Remote call with size payload can introduce additional latency during the inference process. For latency-sensitive applications, this disadvantage might be a deal breaker unless there are mitigations put in place, such as caching.

In summary, the design decision of whether to embed model prediction within the inference service or utilize a separate prediction service depends on many factors, specific use case requirements, and constraints. For scenarios that require high scalability, modularity, diverse and large ML models, and fluctuating prediction traffic volume, a separate prediction service can better support those needs. For scenarios that have a small number of ML models with low prediction traffic, favor simplicity, and have a small team, embedding prediction within the inference service is a reasonable design decision to start with.

It is quite reasonable for a new model serving infrastructure to start with the embedding prediction within the inference approach and then evolve it to the second approach once requirements about scalability, reliability, and large and complex models become important. In fact, the model serving infrastructure at Reddit[3] had evolved in such order, which we will discuss in more details in the Case Studies section.

Offline Inference

Much of the discussion up to this point has been mainly about the necessary infrastructure to support online inference needs. A model serving infrastructure that lacks offline inference is incomplete, since the offline inference plays an important role in maximizing the ROI of ML models. Certain ML use cases need offline inference, requiring infrastructure that can generate offline inferences in a standardized, scalable, and efficient manner. In this section, we will first discuss a few of these use cases before diving into the technical architecture.

Offline inference, also known as batch inference, is the process of periodically generating a vast amount of predictions from a batch of input data. This is in contrast to online inference, which generates predictions from individual data points as they are received. The offline inference process can be either triggered manually on an ad hoc basis or orchestrated by a workflow scheduling system to generate predictions on a regular basis based on time or when new data is available. The generated predictions are typically stored in a data warehouse or data lake for further analysis or consumption by other business processes. In some use cases, these generated predictions are transported to a key-value database to be used.

Examples of ML use cases that need offline inference include

- Churn prediction: Subscription-based service providers, such as telecommunication companies and Netflix, use ML to predict customer churn across their customer base, which could be in millions or hundreds of millions. Offline inference is used to generate the customer churn predictions, enabling the development of strategies to retain their customers.

---

[3] Garrett Hoffman, Evolving Reddit's ML Model Deployment and Serving Architecture, 2021, www.reddit.com/r/RedditEng/comments/q14tsw/evolving_reddits_ml_model_deployment_and_serving/

- Energy consumption forecasting: Energy companies use ML to forecast energy consumption patterns. Offline inference is used to generate predictions of future energy demand based on historical data, previous consumption patterns, and weather conditions.

- Marketing campaign optimization: Marketing teams are constantly looking for ways to optimize marketing campaigns to improve their effectiveness, which leads to increased business values. Offline inference is used to predict campaign effectiveness using historical customer behavior and campaign performance data.

- Satellite image analysis: Various government departments and businesses have an interest in identifying changes in land use, monitoring deforestation and tracking crop growth. Offline inference is used to analyze a large set of satellite imagery to identify changes.

These use cases highlight the need for offline inference across different ML use cases from different industries. Additionally, handling large batches of input data, representing historical data, is crucial during the prediction generation process.

A robust offline inference infrastructure should be capable of generating a large amount of predictions from large datasets, triggered on an ad hoc or scheduled basis, and done so in an efficient manner. To meet these requirements, the core components of such infrastructure include workflow scheduling system and distributed processing engine. Figure 5-6 depicts the high-level design of an offline inference infrastructure, where the key component that enables high throughput is the distributed processing engine.

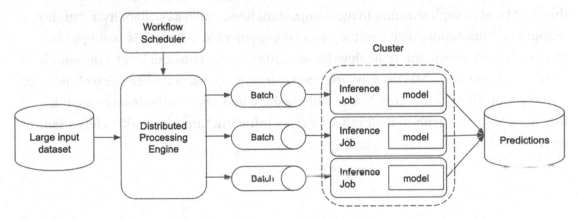

Figure 5-6. *High-level offline inference architecture*

Distributed processing engines like Apache Spark or Ray are great choices for distributing the offline inference workload across a cluster of nodes with powerful compute resources. This approach speeds up the prediction generation and ensures high throughput.

As usual, scaling up offline inference presents numerous interesting challenges. This blog[4] outlines some of the key challenges and provides a comparative analysis of three different solutions for offline image classification inference.

Case Studies

Model serving infrastructure is a key enabler for achieving the ROI on the ML projects. The required level of sophistication a model serving infrastructure needs is driven by a few factors, such as the number of current and future models, the proportion of online and offline inference use cases, the volume of model inference requests and their latency requirements, and the model complexity.

For startups or organizations that are at the beginning of their journey of applying machine learning, it is reasonable to build the first version of their model serving infrastructure with simplicity and minimal effort. Initially, the model deployment might not be fully automated. Additionally, whenever possible, leverage the various components from the open source community and their cloud vendor.

For medium organizations with a growing number of models, scalability, reliability, and efficiency are becoming increasingly important; then it is wise to put a considerable amount of investment and effort to build out their model serving infrastructure with the right level of sophistication to meet important needs such as scalability, reliability, amount of automation, such that the model deployment and serving is as simple as microservice deployment. To achieve these requirements, consider leveraging solutions from cloud providers, MLOps vendors, open source projects, and in-house technologies.

The following section "In-House" will highlight two case studies from companies that have built and evolved their model serving infrastructure, along with a few lessons learned.

---

[4] Amog Kamsetty, Eric Liang, Jules S. Damji, Offline Batch Inference: Comparing Ray, Apache Spark, and SageMaker, 2023, www.anyscale.com/blog/offline-batch-inference-comparing-ray-apache-spark-and-sagemaker

The following section "Open Source" will dive into a few popular open source solutions to consider when building out a new model serving infrastructure or evolving one.

In-House

For companies where ML is deeply embedded in their product offerings, they have a large user base, and they have been operating more than five years, it is very likely that they develop their own in-house model infrastructure to address their specific functional and nonfunctional requirements. This section will highlight two case studies: one from Lyft, a ride sharing company, and the other one from Reddit, a social news and forum company.

LyftLearn Serving

On Lyft ride sharing platform, hundreds of millions of real-time decisions are made each day by machine learning models. These ML-based decisions include pricing optimization, ride allocation optimization, ETA predictions, fraud detection, and more.

The aforementioned ML use cases require online inference at scale. To support these needs and to overcome their legacy model serving infrastructure constraints, the ML platform team at Lyft modernized it with LyftLearn Serving.[5]

LyftLearn Serving is an opinionated model serving infrastructure designed to be robust, performant, decentralized, and designed to meet the requirements listed in Figure 5-7.

---

[5] Hakan Baba, Mihir Mathur, Powering Millions of Real-Time Decisions with LyftLearn Serving, 2023, https://eng.lyft.com/powering-millions-of-real-time-decisions-with-lyftlearn-serving-9bb1f73318dc

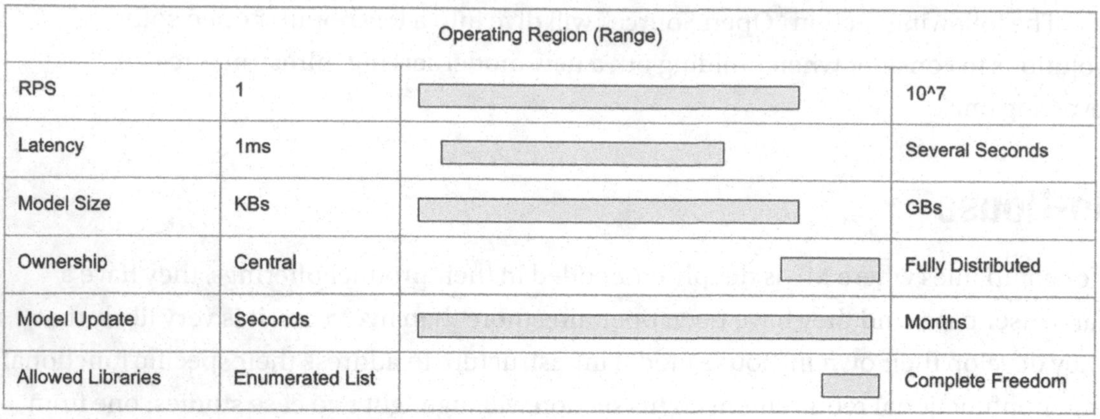

Operating Region (Range)				
RPS	1			10^7
Latency	1ms			Several Seconds
Model Size	KBs			GBs
Ownership	Central			Fully Distributed
Model Update	Seconds			Months
Allowed Libraries	Enumerated List			Complete Freedom

Figure 5-7. *Adapted: LyftLearn Serving requirements. The bar width represents the rough span*

Among the many key design decisions in LyftLearn Serving, two are particularly noteworthy for a detailed examination and learning opportunity: serving library and distributed ownership.

At the heart of the LyftLearn Serving is a library that provides the essential capabilities commonly needed by the model deployment and serving infrastructure. This library contains logics for model loading and unloading, model versioning, model shadowing, inference request handling, model monitoring, model logging, and more. Its modular approach enables seamless integration within microservice architecture. Additionally, this library provides an abstraction for loading models and inference, enabling the implementation of this abstraction to easily inject custom code to meet the specific use case needs. This flexibility enables the LyftLearn Serving library to effectively accommodate the diverse unique needs of various Lyft teams. In terms of the ML library, LyftLearn Serving does not impose any restriction on the framework as long as there is a Python interface for it. Figure 5-8 depicts the high-level inference request flow and shows where in this flow the custom code is executed.

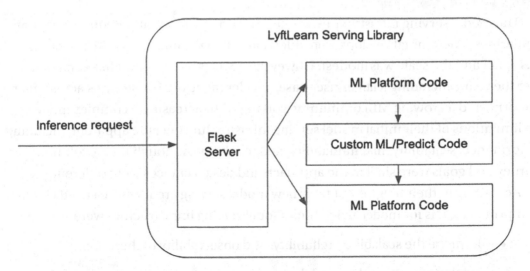

Figure 5-8. *Adapted: LyftLearn Serving Inference request*

One central design consideration of LyftLearn Serving is to empower a decentralized approach to deploying and serving ML models. This decentralized approach provides each team at Lyft with the necessary isolation to operate at their pace and the flexibility to utilize specific ML libraries and versions tailored to their unique needs. Since LyftLearn Serving is essentially a set of libraries rather than services, each team's isolated model serving service can easily incorporate them. In essence, the model deploying and serving through a set of libraries is a key enabler of decentralized approach to ML model deployment and serving.

The concrete isolation mechanism LyftLearn Serving chose is at the GitHub repo level, which helps define clear ownership and boundaries. This means each team at Lyft creates their own repos and each of those repos pulls in the necessary LyftLearn Serving libraries.

In summary, the two key design principles of LyftServing Serving are model serving as a library and distributed ownership model serving. These two principles work in tandem to maximize flexibility and foster isolation, empowering teams at Lyft to make independent decisions regarding their ML library needs and move at their pace for their model serving needs.

Reddit's Model Serving Architecture Evolution

As a social news aggregation and discussion platform, Reddit's ML use cases include personalized feeds, ads targeting, search and discovery, anomaly detection, and more.

The model serving infrastructure evolution journeys at Lyft and Reddit share many similarities. Their initial versions were adequate when the number of ML use cases was small and the scale was modest. However, as ML usage at Reddit has expanded over the years, including a larger user base, a wider range of ML use cases across their platform, and a growing ML community working with increasingly complex models, the limitations of their initial model serving infrastructure became apparent, including performance, scalability, maintainability, and reliability. Although their evolution journey and goals were similar, the approach and design choices were different.

According to their blog[6] about their new model serving architecture called Gazette, their primary goals for modernizing their model serving infrastructure were

- Improve the scalability, reliability, and observability of the system

- Deploy more complex models

- Have the ability and flexibility to better optimize model performance

- Improve the developer experience

The first major architectural change involved splitting the responsibility of handling inference requests, feature fetching, and model prediction into two separate services, as illustrated in Figure 5-9. This separation effectively partitioned the two different workloads: I/O bound and compute bound, into separate services, enabling them to be scaled independently to achieve the aforementioned goals, particularly scalability and reliability.

---

[6] Garrett Hoffman, Evolving Reddit's ML Model Deployment and Serving Architecture, 2023, www.reddit.com/r/RedditEng/comments/q14tsw/evolving_reddits_ml_model_deployment_and_serving/

Reddit Model Serving Architecture

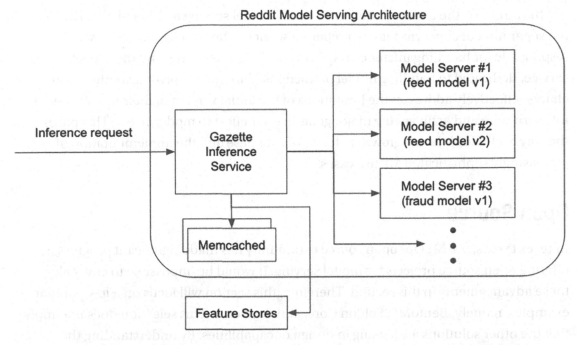

Figure 5-9. *Adapted: Architecture of Gazette inference service and Model Server. Arrows represent the request flow*

The first one is called Gazette inference service, using their Golang web services framework, and is responsible for serving ML inference requests from various clients. This service has a single generic endpoint that all clients send requests to, along with various metadata, such as model name and version. Once the request is received, this service fetches any necessary from the feature stores, and then routes the actual prediction requests to the second service. This new service has better overall performance since it is written in Go programming language, which is much better at handling concurrency.

The second service is called Model Server service, Python based, which wraps a specified ML model. Each instance of the Model Server service is containerized with docker, and can be configured and tailored to specific needs, such as ML library type and version, compute resources, auto scaling parameters, based on the model complexity and expected prediction volume. Since each model is deployed in isolation, this helps achieve the reliability goal, and models are more fault tolerant to unexpected crashes in other models.

In summary, the crux of Reddit's revamped model serving architecture lies in the separation of concerns into two separate services: the Gazette inference service, responsible for handling inference requests and feature fetching, and the Model Server service, dedicated to handling model predictions. This one yet pivotal architectural change effectively addresses the limitations of the initial version of their model serving infrastructure and achieves the major goals in their effort to modernize it. This paves the way for Reddit's future growth in terms of user base and the implementation of increasingly sophisticated ML use cases.

Open Source

In recent years, the MLOps open source community has made significant progress in offering open source projects for model serving. It would be impossible to cover all these advancements in this section. Therefore, this section will focus on a few popular examples, namely, BentoML, Seldon Core, and Ray Serve. This selection does not imply that the other solutions are lacking in design or capabilities. By understanding the strengths and design choices of these popular solutions, you will be better equipped to evaluate solutions that can meet the needs of your company's specific ML use cases and infrastructure requirements.

The following small set of criteria was used to help narrow down a list of open source model serving solutions to highlight:

- Diverse ML framework support: ML domain is fairly broad and evolves rather rapidly. The diverse ML framework support will empower ML modelers with freedom and flexibility to leverage and explore the best ML, ultimately driving business impact.

- Friendly local deployment and testing support: The ML community within each organization will grow as the ML adoption increases. Rapid ML deployment velocity is a key factor to accelerate time to production for ML models; thereby the ROI of ML projects will be realized quicker.

- Scalable and flexible deployment support: ML domain has gained widespread recognition and adoption in the business world. As the adoption continues to grow, scalable and flexible deployment support will play a crucial role in accelerating ROI of ML projects.

These open source solutions are evolving rapidly, so do check their updated website to learn their latest capabilities.

BentoML

According to documentation,[7] BentoML is a unified AI application framework for building reliable, scalable, and cost-efficient AI applications. It provides a comprehensive toolkit for streamlining the model serving, application packaging, and production deployment.

BentoML is a Python-centric framework and has comprehensive support for saving and loading models that are trained from various popular ML frameworks, including TensorFlow, Pytorch, LightGBM, XGBoost, and more. In addition, it provides an easy and flexible integration point to define the custom logic for prediction pre-processing and post-processing while defining the inference service endpoint.

The typical flow for deploying ML models with BentoML includes

- Save trained models into BentoML model store.

- Create BentoML service: Define the definition of each service in a Python file. The service definition includes steps for model loading, runner set up, the inference service endpoint, and any custom logic during the model inference, such as prediction pre-processing and post-processing.

- Build Bento: Bento serves as the standardized distribution unit in the BentoML ecosystem. Each Bento consists of essential components, including source code, models, data files, and dependencies for the service to run. These components are configured in file using the YAML format. BentoML provides command line utility for building Bentos and automatically tagging them with a unique version.

- Once Bentos are built, they can be tested locally or pushed to production environments, including BentoCloud or any Docker-compatible environment, such as Kubernetes, Swarm, Amazon ECS, Google Cloud Run, and more.

---

[7] What is BentoML, https://docs.bentoml.com/en/latest/overview/what-is-bentoml.html

In terms of friendly local deployment, testing, and debugging, BentoML offers several command line utilities to quickly launch the service defined in each Bento locally. Afterward, sample requests can be sent to the service endpoint using HTTP or gRPC protocol to validate both the service and model.

The combination of Service APIs and Runner concepts is how BentoML supports scalable model serving. A BentoML Service serves as the interface for handling inference requests and returned responses. Logically, it encapsulates two main components: Service API and Runner. At runtime, BentoML Services launch two components known as API server and Runner, as depicted in Figure 5-10. Each of these components can be configured to scale independently in terms of compute resource type and the number of instances.

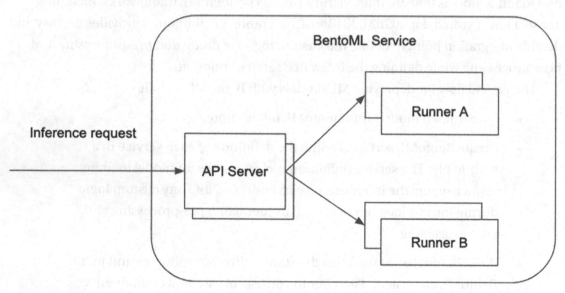

Figure 5-10. *Adapted: BentoML Service architecture*

Each BentoML service can have one or multiple Service APIs, with each representing an endpoint that can be called remotely. Each Service API definition includes a specification of input and output, and a callback function. The callback function is designed to contain the model inference logic, which may include logic like feature fetching, pre-processing and post-processing, and model prediction invocation.

At a high level, the Runner is designed as remote Python workers for performing predictions and can be scaled independently from API servers. Each BentoML Service can be configured to launch a group of Runner workers to meet the expected inference request volume, and the calls to those workers will be distributed accordingly among those Runners. Out of the box, BentoML provides a set of pre-built Runners implemented for each supported ML framework. For advanced use cases where customizations are needed to meet specific requirements, BentoML allows us to create and use our own Runner class.

One built-in feature in Runner to help achieve high throughput during the model prediction generation is called adaptive batching. This is considered as server-side adaptive batching, and this is advantageous as opposed to client-side batching because the inference request clients don't need to incorporate any batching logic on their side.

There are two main concepts[8] in adaptive batching: batching window and batch size.

- Batching window: The maximum amount of wait time to assemble a "batch" of prediction requests before releasing it for processing. This is helpful while the prediction request volume is low.

- Batch size: The maximum number of prediction requests to include in a batch before releasing them for processing. This is useful when the prediction request volume is high.

To illustrate the simplicity of deploying ML models using BentoML, the following section provides various code snippets for saving ML models to BentoML model store, creating a service, building a Bento, and deploying a Bento locally and testing the service endpoint. These code snippets are extracted from the linear_regression[9] example on BentoML GitHub repository.

The code snippet in Listing 5-1 shows how to save a trained ML model to BentoML model store. By default, the model store is located in the bentoml directory under user home directory. The variable BENTOML_HOME can be used to change where the models are stored.

Listing 5-1. Saving a trained ML model to BentoML model store

```
from sklearn import linear_model
import bentoml
reg = linear_model.LinearRegression()
reg.fit([[0, 0], [1, 1], [2, 2]], [0, 1, 2])
bento_model = bentoml.sklearn.save_model("linear_reg", reg)
print(f"Model saved: {bento_model}")
```

---

[8] Adaptive Batching, https://docs.bentoml.com/en/latest/guides/batching.html

[9] BentoML SKlearn Example: Linear Regression, https://github.com/bentoml/BentoML/tree/main/examples/sklearn/linear_regression

The next step is to define a BentoML service in a Python file to serve inference requests and perform reference. Listing 5-2 shows an example of doing that.

Listing 5-2. service.py – define a simple BentoML service

```
import bentoml
from bentoml.io import NumpyNdarray
reg_runner =
    bentoml.sklearn.get("linear_reg:latest").to_runner()
svc =
  bentoml.Service("linear_regression",runners=[reg_runner])
input_spec = NumpyNdarray(dtype="int", shape=(-1, 2))

@svc.api(input=input_spec, output=NumpyNdarray())
async def predict(input_arr):
    return await reg_runner.predict.async_run(input_arr)
```

The preceding code snippet creates a Runner instance and attaches a model loaded from the model store. After that, a BentoML service is initialized with a "linear_regression" name and one runner. The last step is to set up an endpoint by defining a callback function and decorating it with @svc.api decorator. By default, the function name becomes the endpoint URL, which can be easily customized using the route option.

To bring up a BentoML service locally for a simple test, we can use the "bentoml service" command line, and the output looks something like in Listing 5-3.

Listing 5-3. The output from bringing up a local service

```
% bentoml serve service.py:svc
2023-11-25T07:40:04-0800 [INFO] [cli] Environ for worker 0: set CPU thread
count to 8
2023-11-25T07:40:04-0800 [INFO] [cli] Prometheus metrics for HTTP
BentoServer from "service.py:svc" can be accessed at http://localhost:3000/
metrics.
2023-11-25T07:40:10-0800 [INFO] [cli] Starting production HTTP BentoServer
from "service.py:svc" listening on http://0.0.0.0:3000 (Press CTRL+C
to quit)
```

Once the BentoML service `linear_regression` is running and listening on port 3000, we can send a simple inference request to the `predict` endpoint using curl command. An example of the command and output is shown in Listing 5-4.

Listing 5-4. An inference request and the response

```
% curl -X POST -H "content-type: application/json" --data "[[5, 3]]"
http://127.0.0.1:3000/predict

[3.9999999999999996]  # response
```

To build a Bento for service `linear_regression`, the first step is to define the configuration of the essential components in a `bentofile.yaml` file, and then run the command `bentoml build` to actually build the Bento. These steps are left as an exercise for the reader.

For illustration purposes, the above example is a simple one. Getting started with BentoML is straightforward. For advanced usage of BentoML, be sure to explore the "Advanced Guides" and "Best Practices" sections of its documentation website at `https://docs.bentoml.com/`.

Seldon Core

At the highest level, Seldon Core[10] is an opinionated, flexible, comprehensive, and sophisticated open source platform to package, deploy, monitor, and manage machine models on top of Kubernetes at a large scale.

It is opinionated in terms of Kubernetes-native. Seldon Core is designed to work seamlessly with Kubernetes and leverages its scalability and flexibility for deploying models. Given that Kubernetes has become the de facto container and proven orchestration platform for managing containerized workloads and services at scale, this is not necessarily a bad decision.

It is comprehensive because it provides advanced model serving capabilities, including inference graph, advanced metrics, request logging, explainers, outlier detection, A/B tests, canaries, and more.

Seldon Core deploys ML models in production as microservices that can handle REST or gRPC requests. It is framework agnostic and has built-in support for commonly used ML frameworks, such as TensorFlow, PyTorch, XGBoost, LightGBM, Scikit-Learn, and more.

---

[10] Seldon Core, `https://docs.seldon.io/projects/seldon-core/en/latest/index.html`

In terms of working with Seldon Core and making the best use of its capabilities, it is essential to be familiar with these main concepts: Model, Server, Pipeline, and Experiment. Figure 5-11 illustrates these concepts and their corresponding concrete capabilities.[11] The following sections will briefly cover the details of each concept, starting from the left-hand side.

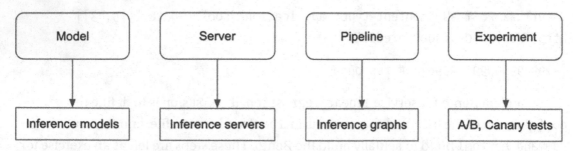

Figure 5-11. *Adapted: Mapping concepts to concrete capabilities*

Models provide the atomic building blocks of Seldon. Examples of models include ML model, drift detectors, outlier detectors, feature transformation, and more. Once models are defined in YAML format, then we can use the provided command line utility to load them onto the Seldon Core cluster and start making inference requests. Listing 5-5 shows a simple example of a Scikit-Learn iris classification model, and Listing 5-6 shows the commands to load the model and make inference call. For Seldon Core Kubernetes deployment, the commands will be slightly different and mainly using the kubectl command.

Listing 5-5. A model definition of Scikit-Learn ML model in sklearn-iris.yaml

```
apiVersion: mlops.seldon.io/v1alpha1
kind: Model
metadata:
  name: iris
spec:
  storageUri: "gs://seldon-models/scv2/samples/mlserver_1.3.5/iris-sklearn"
  requirements:
  - sklearn
  memory: 100Ki
```

---

[11] Clive Cox, Ed Shee, Launch of Core V2, www.seldon.io/webinar/launch-of-core-v2

Listing 5-6. Command lines to load and make inference call

```
# load the model onto Seldon Core platform
seldon model load -f ./models/sklearn-iris.yaml
# check the status
seldon model status iris -w ModelAvailable | jq -M .
# make REST inference call
seldon model infer iris \
  '{"inputs": [{"name": "predict", "shape": [1, 4], "datatype": "FP32",
  "data": [[1, 2, 3, 4]]}]}'
```

Once there are few defined models, we can leverage the pipelines capability to connect them into flows of data transformation, also known as inference graphs. This capability allows us to form complex computational graphs that represent the structure and flow of operations during the mode inference. Pipelines are a powerful tool for optimizing and accelerating the execution of complex ML inference needs, particularly in production environments where real-time performance and scalability are essential. In addition to chaining the output of one model into the input of another, pipelines also support combining outputs from multiple models as an input to another model, as well as conditional flows. For more details about the various sophisticated pipelines features, please refer to its documentation at `https://docs.seldon.io/projects/seldon-core/en/v2/contents/pipelines/index.html`.

Canary testing as a part of software deployment is a good practice in DevOps. The similar technique is quite applicable when deploying ML models to production. An Experiment in Seldon Core can support traffic splitting between model and pipelines and traffic mirroring. Listing 5-7 shows a simple example of a split of 50% traffic between two models `iris` and `iris2`.

Listing 5-7. An experiment of splitting traffic evenly across two models

```
apiVersion: mlops.seldon.io/v1alpha1
kind: Experiment
metadata:
  name: experiment-sample
spec:
  default: iris
  candidates:
```

```
- name: iris
    weight: 50
- name: iris2
    weight: 50
```

Test models in a shadow mode is just as simple using the mirroring support. See Listing 5-8 for a simple example of testing a model named iris2 in shadow mode with 100% traffic.

Listing 5-8. An experiment of mirroring traffic to test iris2 in shadow mode

```
apiVersion: mlops.seldon.io/v1alpha1
kind: Experiment
metadata:
  name: sklearn-mirror
spec:
  default: iris
  candidates:
  - name: iris
      weight: 100
  mirror:
      name: iris2
      percent: 100
```

Seldon Core has a sophisticated architecture,[12] as depicted in Figure 5-12, designed to support a comprehensive set of capabilities mentioned above and to support model serving at scale. The core components can run locally via Docker Compose or be deployed onto Kubernetes clusters.

---

[12] Seldon Core V2 Architecture, https://docs.seldon.io/projects/seldon-core/en/v2/contents/architecture/index.html

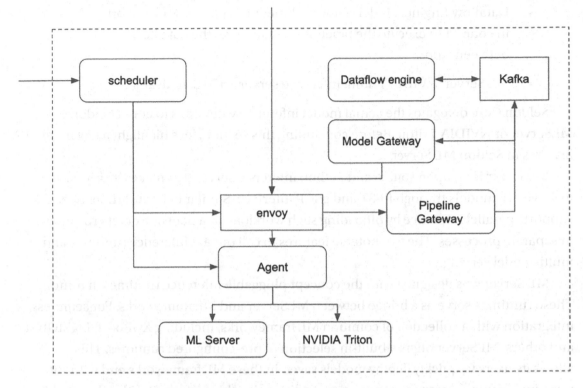

Figure 5-12. *Adapted: Seldon Core V2 architecture*

The scheduler component accepts requests from various clients, such as orchestration systems, Kubernetes, or other external services, to manage the deployment and scaling of ML models.

The envoy component accepts requests from various clients, such as external services, API gateway, or orchestration systems, to direct traffic to deployed models for predictions.

The core responsibilities of each component are described as follows:

- Scheduler: Acts as the main entry point for interacting with the Seldon Core platform. It manages the loading and unloading of models, pipelines, experiments, and more.

- Envoy: Serves as proxy to load balance and route incoming prediction requests.

- Agent: Manages the interactions with inference servers, ML Server and NVIDIA Triton, including inference request routing, model loading and unloading, and more.

- Dataflow Engine, Model Gateway, Pipeline Gateway, Kafka: Work in concert to execute the pipelines and manage the data flow between steps.

- ML Server, NVIDIA Triton: Inference server for ML models.

Seldon Core delegates the actual model inference workloads to either Seldon MLServer or NVIDIA Triton. Before concluding this section, let's highlight a few notable features in Seldon ML Server.

MLServer is an open source and Python inference server that makes it very easy to serve ML models through REST and gRCP interface. Similar to BentoML service, it supports parallel inference by offloading such workload to a pool of workers running in separate processes. The two notable features to call out are inference runtimes and multi-model serving.

ML Server was designed with the concept pluggable inference runtimes in mind. These runtimes serve as a bridge between MLServer and ML frameworks. For seamless integration with a collection of common ML frameworks, including XGBoost, LightGBM, and others, MLServer offers a built-in selection of pre-configured runtimes. This allows its users to quickly deploy models saved in these ML frameworks and without introducing unnecessary dependencies. Additionally, this feature provides the flexibility of integrating with custom runtimes if needed. Interestingly enough, there are no built-in inference runtimes for TensorFlow and PyTorch ML frameworks. To learn about the most up-to-date list of included inference runtimes, visit this resource at `https://mlserver.readthedocs.io/en/latest/runtimes/index.html`.

MLServer's built-in multi-model serving support enables a single instance of MLServer to serve multiple models and their respective versions. For organizations with a large and growing number of models, this feature promotes cost saving by maximizing the compute usage while keeping a minimal infrastructure footprint. Figure 5-13 visually contrasts the two model serving approaches, highlighting the benefits of multi-model serving. The multi-model serving option on the right-hand side shows that we can use an optimization technique to efficiently place multiple models together on a single deployment instance, maximizing resource utilization and consequently reducing the overall required resources in terms of memory, CPUs, or GPUs.

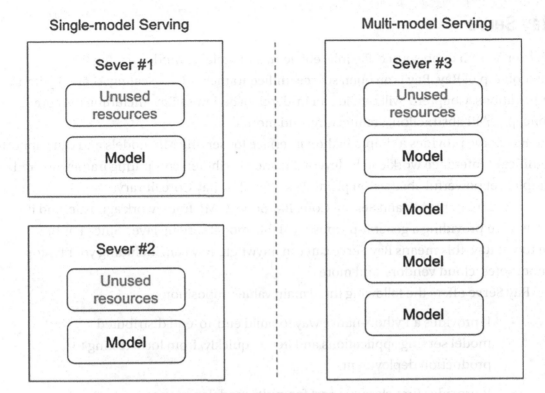

Figure 5-13. *Single-model vs. multi-model serving*

However, it is important to carefully consider which models should share the same server to avoid potential "noisy neighbor" problems, where resource demands of one model, such as memory or CPU, can negatively impact the performance of others.

Seldon Core is an impressive model serving platform with a comprehensive set of features to support enterprise-level model serving needs. It is designed for mid- to large-sized organizations that have a large model serving workload and with a growing number of ML models trained using a diverse set of ML frameworks. Seldon Core is designed to work closely with Kubernetes to leverage its scalability and orchestration capabilities.

Ray Serve

Ray Serve is an open source, flexible, efficient, and scalable model serving library built on top of Ray. Ray is an open source unified framework for scaling AI and Python applications. Chapter 7 will provide an in-depth overview of Ray, including the core concepts, API abstraction, architecture, and more.

Ray Serve provides a simple Python interface for serving ML models and can support significant inference workload by leveraging the distributed computing framework and simple, yet powerful abstraction primitives offered by Ray Core library.

Similar to BentoML and Seldon Core, Ray Serve is ML framework agnostic, and it focuses on providing a general-purpose scalable model serving layer. Since it is built on top of Ray, this means Ray Serve can run anywhere Ray can, including your laptop, Kubernetes, cloud vendors, and more.

Ray Serve offers the following three main value propositions:

- It provides a Python-native way to build end-to-end distributed model serving applications and iterate quickly, from local testing to production deployment.

- It provides first-class support for multi-model inference.

- It provides flexible scaling and resource allocation.

Before delving into each of the value propositions, let's discuss a few central concepts in Ray Serve: deployment and application.

A deployment is essentially a deployable model serving unit that can contain any kind of logic written in Python, including business logic, model loading and inference logic, feature fetching logic, and inference pre-processing and post-processing. To define a deployment, decorate a Python class definition with a @serve.deployment decorator. At runtime, one or multiple copies of the class are started in separate processes running in a Ray Serve cluster. The number of processes can be scaled up or down or autoscaled to match the incoming request load.

An application consists of one or more deployments and acts as the unit of upgrade in a Ray Serve cluster. Typically a set of deployments in application work together to meet the needs of a model serving application, such as image processing or audio transcription.

Python-Native Model Serving Application

Ray Serve offers unparalleled flexibility for developing model serving applications in a Python-centric way. A Python class serves as your canvas for expressing the model serving logic, ranging from simple code snippets to intricate workflows spanning thousand lines of code. This flexibility comes with responsibility for dealing with loading models in certain formats and how to work with different ML frameworks.

All that is needed to turn a Python class into a Ray Serve application is by adding a simple, yet powerful decorator @serve.deployment, and then Ray Service will take care of creating multiple copies of the Python class and deploying them into the Ray Service cluster. Behind the scenes, Ray Serve applications are managed by a centralized controller, which is responsible for creating and managing Ray Serve applications, such as failure detection and recovery.

Listing 5-9 is a very simple Ray Serve application adapted from an example available in Ray GitHub.[13] In this example, the constructor loads the "sentiment-analysis" model from the Hugging Face model hub. The __call__ method will be called to handle the incoming requests, and it expects a query parameter name called "text."

Listing 5-9. An experiment of mirroring traffic to test iris2 in shadow mode

```
import requests
from starlette.requests import Request
from typing import Dict
from transformers import pipeline
from ray import serve

# 1: Wrap the pretrained model in a Serve deployment.
@serve.deployment
class SentimentAnalysisDeployment:
    def __init__(self):
        self._model = pipeline("sentiment-analysis")

    def __call__(self, request: Request) -> Dict:
        return self._model(request.query_params["text"])[0]

# the bind() API instantiates a copy of this class
sentiment_app = SentimentAnalysisDeployment.bind()
```

---

[13] Ray GitHub repository, https://github.com/ray-project/ray.git

Ray Server provides two different ways to launch a deployment. The first way is done by calling the serve.run Python API, and the second way is done by using the serve run CLI command (make sure to install Ray Serve first). Listing 5-10 shows the results from running the CLI with the assumption of the preceding code in a file called transformer.py.

Listing 5-10. Launch a deployment using CLI and show the output.

```
% serve run transformer:sentiment_app
# output from launching transformer:sentiment_app deployment
2023-11-26 17:42:48,480    INFO scripts.py:501 -- Running import path:
'transformer:sentiment_app'.
2023-11-26 17:42:54,892    INFO worker.py:1664 -- Started a local Ray
instance. View the dashboard at 127.0.0.1:8265
(ProxyActor pid=29381) INFO 2023-11-26 17:43:00,183 proxy 127.0.0.1 proxy.
py:1072 - Proxy actor dd1aa8088598b2d9635fdfdb01000000 starting on node
64a62a755106ef22993a76e3737bd5cd5218e82444a602f6a88b4580.
(ProxyActor pid=29381) INFO 2023-11-26 17:43:00,194 proxy
127.0.0.1 proxy.py:1257 - Starting HTTP server on node:
64a62a755106ef22993a76e3737bd5cd5218e82444a602f6a88b4580 listening on
port 8000
(ProxyActor pid=29381) INFO:    Started server process [29381]
(ServeController pid=29380) INFO 2023-11-26 17:43:00,356 controller
29380 deployment_state.py:1379 - Deploying new version of deployment
SentimentAnalysisDeployment in application 'default'.
(ServeController pid=29380) INFO 2023-11-26 17:43:00,459 controller
29380 deployment_state.py:1668 - Adding 1 replica to deployment
SentimentAnalysisDeployment in application 'default'.
```

Once the deployment is up and running, now we can start sending inference requests by making REST calls to http://localhost:8000/ endpoint. Listing 5-11 shows several REST calls and their responses.

Listing 5-11. Sending inference requests via REST calls

```
% curl -G "http://localhost:8000/" --data-urlencode "text=the world is
falling"
{"label": "NEGATIVE", "score": 0.9988685846328735}
% curl -G "http://localhost:8000/" --data-urlencode "text=today is a
beautiful day"
{"label": "POSITIVE", "score": 0.9998778104782104}

% curl -G "http://localhost:8000/" --data-urlencode "text=tomorrow
is Monday"
{"label": "POSITIVE", "score": 0.9901313781738281}

% curl -G "http://localhost:8000/" --data-urlencode "text=tomorrow is
Monday and I have to work"
{"label": "POSITIVE", "score": 0.9546266794204712}

% curl -G "http://localhost:8000/" --data-urlencode "text=tomorrow is
Monday and I have to work again"
{"label": "NEGATIVE", "score": 0.9834485650062561}
```

Flexible Scaling and Resource Allocation

ML model inference is a computationally intensive task. Make sure to allocate a sufficient amount of resources to your model serving applications so they can handle the expected inference workload.

Ray Serve provides two tuning knobs to scale deployments and use appropriate compute resources, such as CPUs or GPUs. These knobs can be specified in the @serve. deployment decorator using these parameters: num_replicas and ray_actor_options. Listing 5-12 shows an example of configuring the sentiment_app with 3 replicas and 2 GPUs.

Listing 5-12. Scale deployment up with multiple replicas and GPUs

```
@serve.deployment(
    num_replicas=3, ray_actor_options={"num_gpus":2})
class SentimentAnalysisDeployment:
    def __init__(self):
```

```
        self._model = pipeline("sentiment-analysis")

    def __call__(self, request: Request) -> Dict:
            return self._model(request.query_params["text"])[0]
```

The above scaling example uses a manual scaling approach. We can leverage the auto scaling feature to react to traffic spikes by automatically scaling up the number of replicas. Ray Serve has an autoscaler component, which monitors the request queue sizes and uses that information to adjust the number of needed replicas. The autoscaling configurations can be specified using the `autoscaling_config` parameter in the `@serve.deployment` decorator. For a complete list of options, please visit the AutoscalingConfig documentation page at `https://docs.ray.io/en/latest/serve/api/doc/ray.serve.config.AutoscalingConfig.html`.

Multi-model Inference

Complex ML model serving applications typically require multiple models to work in concert to carry about a specific application goal. Examples of complex ML model serving applications include audio transcription, computer vision, text summarization, and more. Let's take an example of a hypothetical audio transcription application that has three steps, visually depicted in Figure 5-14: audio preprocessing, speech-to-text, and proofreading for accuracy and consistency. Each of these steps will use a different model, each model has different compute resource requirements, such as GPUs, and each step needs to auto scale independently.

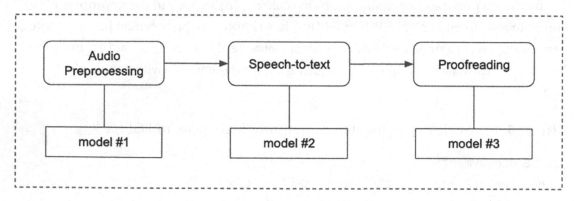

Figure 5-14. Multi-step audio transcription ML model serving application

Ray Serve provides a simple and elegant approach of composing multiple deployments together by passing deployment references between them. The example above demonstrates a pipeline pattern, where the output of one deployment is fed as input to the next. In addition to the pipelining pattern, this blog[14] describes the other three patterns and shows how to implement them using Ray Serve.

Listing 5-13 shows a code snippet that wires three deployments pipelining manner, mirroring the steps in the hypothetical audio transcription application depicted in Figure 5-14.

Listing 5-13. Code skeleton to set up a pipeline of deployments

```
%@serve.deployment
class AudioPreprocessing:

@serve.deployment
class SpeechToText:

@serve.deployment
class ProofReading:

audio_app = AudioPreprocessing.bind()
speech_to_text_app = SpeechToText.bind(audio_app)
proof_reading_app = ProofReading.bind(speech_to_text_app)
```

Multi-model inference capability in Ray Service is an extremely powerful tool that not only simplifies production ML pipelines, but also contributes to cost reduction. After introducing Ray Server at Samara, they saw a ~50% reduction in total ML inference cost per year for their company.[15]

Two other notable Ray Serve features worth mentioning here are model multiplexing and dynamic request batching, and their details can be found at https://docs.ray.io/en/latest/serve.

---

[14] Simon Mo, Edward Oakes, Michael Galarnyk, Serving ML Models in Production: Common Patterns, 2021, www.anyscale.com/blog/serving-ml-models-in-production-common-patterns

[15] Pang Wu, Building a Modern Machine Learning with Ray, 2023, https://medium.com/samsara-engineering/building-a-modern-machine-learning-platform-with-ray-eb0271f9cbcf

Summary

Model serving infrastructure is one of the key pillars of machine learning infrastructure, enabling organizations to capitalize on the ROI of ML projects by facilitating model inference in production for various impactful ML products, such as Ads targeting, personalized recommendation, fraudulent detection, and more. Compared to the feature engineering and model training infrastructures, it is vast and complex, and may require considerable software engineering expertise to support large-scale real-time inference use cases with diverse compute resource needs.

The sophistication of model serving infrastructure level is determined by a few factors, including the number of current and anticipated models, the proportion of online and offline inference use cases, the volume of model inference requests and their latency requirements, and the complexity of models themselves. A well-developed model serving infrastructure needs to seamlessly integrate with the following components and infrastructures:

- Model registry: Load ML models from

- Feature store: To retrieve features from

- Logging infrastructure: To send prediction logs to

- Metrics infrastructure: To send operational metrics to

In the last few years, numerous viable model serving solutions are available from the open source community, such as BentoML, Seldon Core, Ray Serve, and more. In general, these solutions have the following commonalities:

- They provide abstractions to facilitate model serving needs, and their users only need to focus on the serving logic.

- They provide an easy path to deploy mode serving infrastructure onto widely adopted Kubernetes platform.

- They provide support for commonly needed features like adaptive batching, inference graph, also known as model composition, and more.

- They provide ways to allocate the needed compute resources.

- They provide integration paths to sending prediction logs and metrics to commonly adopted logging and metrics infrastructures.

For organizations with a small number of real-time inference machine learning use cases, utilizing vendor-managed model serving solutions can be a viable option. When the number of use cases reaches a certain threshold, such as twenty or more, and when the scalability and stringent latency requirements dominate the conversations, then consider building an in-house model serving infrastructure on top of one of the open source solutions or their equivalent cloud-based version.

CHAPTER 6

ML Observability Infrastructure

Congratulations! After weeks or months training, your model is finally complete and deployed to production, and integrated into your recommendation system. The A/B testing results show a meaningful and positive impact on business metrics. However, your co-worker reminds you that your work is not yet finished. Your model needs continuous monitoring to ensure its performance remains optimal. In other words, your model is at the beginning of its operational journey.

It is a well-known fact that the performance of ML models degrades over time once they are deployed to production. As such, their performance needs to be continuously monitored and observed to detect issues and to prevent negative consequences, such as business impacts or degraded user experience. There are many reasons behind such a phenomenon. At the center of it is that both trained models and the ones in production depend on data. As we know, the world doesn't stay constant; the social and cultural norms evolve over time. Sometimes there are sudden and dramatic shifts, such as the recent COVID-19 pandemic, which had a profound impact on the behavior of people around the world. Models trained on pre-pandemic data struggled to adapt to the "new normal," leading to decreased accuracy and reliability. For example, Lyft's estimated times of arrival (ETAs) model over-predicted ride times,[1] which caused a chain reaction of poor performance in models that used ETAs as an input.

---

[1] Mihir Mathur and Jonas Timmermann, Full-Spectrum ML Model Monitoring at Lyft, 2022, https://eng.lyft.com/full-spectrum-ml-model-monitoring-at-lyft-a4cdaf828e8f

© Hien Luu, Max Pumperla and Zhe Zhang 2024
H. Luu et al., *MLOps with Ray*, https://doi.org/10.1007/979-8-8688-0376-5_6

Adding another layer of complexity on top of the gradual model performance degradation is the phenomenon of silent failures. Unlike traditional software errors that trigger explicit warnings or errors like "page not found," models can appear to function normally while their performance gradually deteriorates. This degradation can reach the point of inaccurate predictions without raising clear warnings or failures to indicate there is a problem. Therefore, it might take weeks before the impact on business metrics becomes noticeable. Therefore, as with any complex system, it is necessary to continually inspect, maintain, and update models to ensure the long-term success and reliability of any ML-powered system.

In essence, it is not wise to "deploy and forget" ML models. Organizations should invest in ML observability to mitigate business risks, such as reputation damage, decreased customer satisfaction, and revenue loss. While it adds extra engineering cost, this investment pays dividends in the long run, particularly for organizations where machine learning is deeply and broadly embedded across their product offering.

ML observability encompasses the tools, practices, and processes to monitor, measure, and understand ML systems, including the performance, feature quality, behavior, and health of ML models and the entire ML pipelines.

Often, the phrase ML monitoring is used interchangeably with ML observability. While they are closely related and are crucial for managing and maintaining models in production, their scopes and functionalities differ. ML monitoring focuses on tracking ML model performance by collecting and monitoring key metrics, such as accuracy, precision, recall, data drift, model drift, and other relevant metrics. It primarily addresses the "what" aspect and serves as a reactive measure to detect model performance or behavior-related problems. ML observability takes a broader approach, expanding beyond ML monitoring to understand and gain insights into the entire ML system and workflow, including model, data, training, pipeline, and deployment environment. It aims to answer the "why" and "how" questions, providing a holistic view of the system's behavior, performance, and potential issues to facilitate troubleshooting and root cause analysis and enable model owners to make informed decisions about model improvement and optimizations.

In short, ML monitoring is a subset of the ML observability strategy. An image of an iceberg, as depicted in Figure 6-1, is commonly used to describe ML monitoring and ML observability. The visible part or the tip of the iceberg represents ML monitoring. The huge portion of the iceberg hidden under the water line represents all the blindspots – data issues, configuration problems, bias, deployment environment, data drift, and concept drift. ML observability aims to expose the entire iceberg, not just the visible tip.

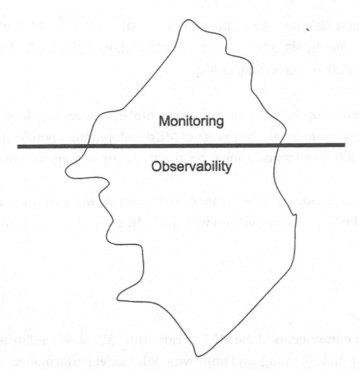

Figure 6-1. *An iceberg metaphor to describe ML monitoring and ML observability*

Note   Monitoring and observability

From the perspective of understanding system behavior, monitoring and observability are related but distinct concepts. As described in the paper "Towards Observability Data Management at Scale,"[2] monitoring refers to collecting metrics and measurements to track system health and performance, such as uptime, latency, errors, and so on. Essentially it focuses on "what" is happening within the system. Observability expands on monitoring by providing contextual insights into the operational state of the whole system to facilitate proactive troubleshooting of hidden issues. Effective observability requires a shift from managing individual metrics to managing the relationships between them to gain a deeper understanding of the system's internal dynamics, interactions, and hidden

---

[2] Suman Karumuri, Franco Solleza, Stan Zdonick, and Nesime Tatbul, Towards Observability Data Management at Scale, 2021, https://dl.acm.org/doi/10.1145/3456859.3456863

complexities, and enabling the identification of root causes. This requires tracing flows across components, aggregating and correlating metrics with logs, and a more comprehensive and holistic analysis.

ML observability infrastructure acts as a watchful eye, providing deep insights into the model performance at every stage of ML development lifecycle, from feature generation to prediction, empowering proactive problem-solving and continuous improvement.

The upcoming sections will delve into the architecture of this infrastructure. Following that, the Case Studies section will highlight a few in-house as well as open sources solutions.

Overview

As one of the key components of the ML infrastructure, ML observability is the backbone for monitoring, troubleshooting, and improving ML model performance. It provides end-to-end visibility into the entire ML development lifecycle, from feature generation, model training and deployment, model predictions, and more. It includes a collection of tools, practices, and technologies to gain deep insights into the behavior and performance of features, models, and ML systems. This enables teams to proactively and quickly discover and mitigate issues that appear during the ML lifecycle before they impact business KPIs. This proactive approach is fueled by these benefits:

- Early detection: Identify and address model performance degradation, bias, and data, concept drift, and others early on.

- Root cause analysis: Delve into the underlying causes of model behavior, from incoming data to model deployment, to understand why it's making certain predictions.

- Enhanced collaboration: Foster a share understanding of model health and predictions among stakeholders, data scientists, and ML engineers, bridging the gap between different teams.

- Continuous improvements: Drive the continuous model improvements through a tight feedback loop of continuously monitored models and actionable insights.

By unlocking these benefits, ML observability becomes the cornerstone of trustworthy and reliable ML models in production.

This section covers the high-level architecture and dives into the essential components and functionalities of ML observability infrastructure. Understanding these technical details will be valuable to building in-house solutions with open source technologies, evaluating vendor solutions or adopting a hybrid approach.

At its core, the ML observability infrastructure needs to provide monitoring, alerting, and analysis capabilities for model performance, drift, data, and explainability, as depicted in Figure 6-2.

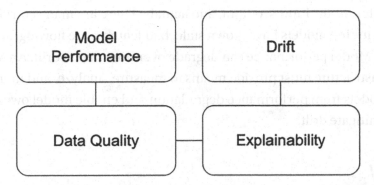

Figure 6-2. *Four areas of analysis and monitoring in ML observability*

Model Performance

A common way to understand the model performance is by measuring and examining the key set of metrics, including accuracy, precision, recall, and more. The performance analysis ensures the model performance has not degraded significantly from when it was trained or when it was initially deployed to production. An ML observability infrastructure must closely track the key metrics against established baselines and trigger alerts when performance dips below acceptable thresholds.

Note   Model performance – fairness, biases, and integrity

Monitoring model performance is crucial for ensuring fairness, addressing biases, and upholding model integrity in machine learning. By regularly evaluating the model's performance on different demographic groups and edge cases, this can help identify any disparities or issues that may arise. Additionally, model performance monitoring can help detect any anomalies or malfunctions that may occur over time, allowing for prompt intervention and maintenance to preserve the model's integrity and reliability.

Drift

The drift concept in ML comes from statistics discipline. At the high level, it refers to the change in the statistical properties of data over time. It is important to measure, track, and monitor drift for model inputs, output, and actuals. There are many contributing factors to drift, including models have grown stale, bad features are flowing, adversarial inputs, and more. Model performance can degrade over time due to drift. An ML observability infrastructure must provide means to measure, analyze, and monitor drift to help protect models from performance degradation and enable model owners to understand and mitigate drift.

Data Quality

The phrase "garbage in, garbage out" perfectly captures how model performance is directly dependent on the quality of input data, both during training and prediction. We all know reality isn't static, and data reflects this. A common set of data quality issues that are relevant to model performance include cardinality shifts, missing data, data type mismatch, out-of-range, and more. Consequently, any robust ML observability infrastructure must actively detect and track these data quality issues throughout the model development lifecycle.

Explainability

Explainability is a tool that helps demystify model decisions. It enables us to trace the reasoning behind a prediction, revealing the key features and their influence on the outcome through feature importance or attribution features. Two popular methods for generating explainability metrics are SHAP and LIME. Integrating explainability alongside the other mentioned capabilities within an ML observability infrastructure proves invaluable for diagnosing root cause and gaining insights into how the models make predictions.

Note   Feature drift vs. model drift

Feature drift refers to the changes in the statistical properties of the input features in a ML model. Specifically, this refers to the feature characteristics, such as average values, standard division, range, and shift compared to the base line. Possible causes of feature drift include seasonal changes in data patterns, changes to the way the data collection methods, external factors like economic crises, demographic changes, and more.

Model drift, also known as concept drift, refers to the deterioration of a model's performance over time, while the feature distribution stays the same. This deterioration is due to a change in the relationship between the input features and the outcome a model is trying to predict. Possible causes of model drift include user behavior changes over time due to new technologies, economic conditions, regulatory changes, and more. Similar to feature drift, model drift compromises the model's accuracy and reliability. An example of model drift is a sentiment analysis model deployed for social media monitoring encountered model drift as the language and expressions used by users evolved over time, leading to a decrease in the model's accuracy in capturing current sentiments.

High-Level Architecture

An effective and robust ML observability architecture should have a set of components to supply the capabilities mentioned above. The high-level architecture is depicted in Figure 6-3, where it places the observability store as the main component for storing model-related data and accessing the aggregated and transformed metrics for ML Observability purposes. The various model-related data comes from feature engineering, model training, and model prediction, and those raw and granular pieces of data will be aggregated and transformed for the purpose of monitoring and alerting, model performance analysis, and model explainability.

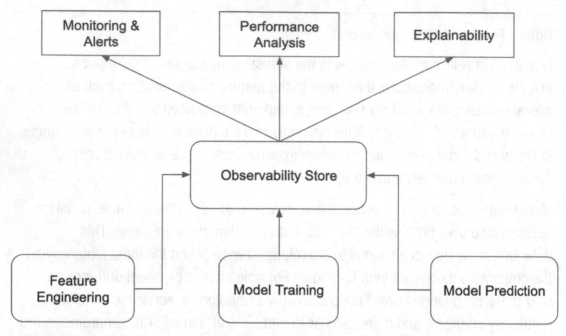

Figure 6-3. *ML observability infrastructure high-level architecture*

Similar to standard microservice observability, ML observability must also incorporate the three main pillars of system observability: logs, metrics, and tracing.[3] The telemetry data will provide additional context about the surrounding systems while interacting with model inference service. This context will be valuable for ML performance tracing, which is the practice of tracing or pinpointing the source model performance issues and mapping back to the underlying data or system issues. By leveraging the rich telemetry data, ML performance tracing provides a clear understanding of the entire inference request lifecycle, identifying the where and why of performance degradation or errors, whether they stem from data issues, infrastructure hiccups, or changes in input patterns.

---

[3] Samuel James, The 3 Pillars of System Observability: Logs, Metrics, and Tracing, 2020, https://iamondemand.com/blog/the-3-pillars-of-system-observability-logs-metrics-and-tracing/

Observability Store

Unlike the feature store and model store, where each of those stores holds one specific type of information for a specific purpose, the observability store is designed to hold a diverse type of model-related data that are produced across the stages of the ML development process, including feature engineering, model training, model evaluation, and model prediction. This assortment of data is what makes the observability store unique and valuable and to meet the needs of ML observability: quickly surfacing up model performance issues and reducing time to resolving those issues.

The following sections briefly outline what data should be logged in the observability store at each of those three phases of ML development lifecycle.

Feature Engineering

Fresh and high-quality data fuel model performance. Production models powered by real-world data suffer from stale or inaccurate features, causing suboptimal outcomes. To proactively combat this, integrate data quality analysis and feature distribution computation as the final step in the feature generation pipelines. These insights should be imported into the observability store to enable quick diagnosis of model performance issues stemming from data quality or feature drift.

Model Training

At the end of a model training job, be it pre-production model training or ongoing model continuous training, it is crucial to log the data distribution, and the overall and sliced model performance metrics with observability store. Capturing information and establishing them as the baselines will facilitate precise performance tracing in case of future issues.

Model Prediction

During the inference process, while production models are used to make predictions, numerous pieces of valuable data aid in monitoring and diagnosing their performance. This data should be logged in the observability store. Examples include input features, actual predictions, production model name and version, and more. Furthermore, operational metrics enrich this data – prediction latency, any errors, inference request client, etc., providing a holistic view for proactive troubleshooting and performance tracing.

225

Observability Store Implementation

The observability store is a concept that is inspired by Josh Tobin[4]'s evaluation store idea. It takes a step further to capture all the relevant ML observability data in a centralized place to meet the needs of monitoring and alerting, drift detection, performance analysis and tracing, and explainability. There have not been any publicly available implementations of the observability store. As for the ML observability vendors, they must have a variation implementation that is tailored to the functionalities their product provides and to distinguish themselves.

When considering coming up with a possible implementation of the observability store, we should first establish a good understanding of the data and metrics that are stored, accessed, and displayed, and then work backward from there to come up with a workable technical design.

The central data entity in interest is the model. Each model has an associated set of performance metrics from training, evaluation, and production. Additionally, each model has an associated set of features that were used to train, evaluate, and perform predictions on. There will be metadata about the prediction request clients, monitoring and alert thresholds, and more.

One key piece of data for understanding and monitoring model performance is the ground truth, the actual right answer. Comparing predictions with the ground truth allows us to truly assess the model's performance level. In some use cases, such as ads targeting or food delivery ETAs, the ground truth becomes available shortly after the predictions are made. However, for others, like predicting fraudulent activities or disease progression, the ground truth could have significant delays, potentially reaching months.

Access patterns generally center on analyzing model performance metrics or conducting drift detection over specific data slices (cohorts) within specified time windows. This analysis can focus on either features or the model itself, depending on the specific use cases. For certain scenarios, such as assessing food delivery ETA accuracy, the ability to analyze model performance and detect drift in near real-time is critically important. Specific examples of the access patterns can be illustrated through the following examples:

---

[4]Josh Tobin, A Missing Link in the ML Infrastructure Stack, 2021, http://josh-tobin.com/assets/pdf/missing_link_in_mlops_infra_031121.pdf

- What is the average drift of all the features of model A in the last 1 hour?

- What is the average drift of model B predictions in the last 24 hours?

- What is the accuracy of the ETA prediction model for state California in the last 4 hours?

Beyond the core capabilities mentioned above, for organizations with a large number of use cases and high volume traffic, the implementation must be able to support a large number of models, metrics, and predictions without sacrificing the latency for the various access patterns mentioned above.

The high-level implementation of an observability store, depicted in Figure 6-4, advocates for the usage of an online analytical processing (OLAP) system and metadata store. The data, metrics, predictions, and ground truth stored in the data lake are organized, aggregated, and transformed using a distributed data processing system, such as Spark, and then continuously ingested into the OLAP system and metadata of the observability store.

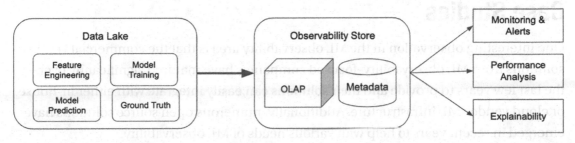

Figure 6-4. *High-level implementation of observability store*

Once the ML observability data is organized in a star schema structure, the OLAP system excels at low-latency analytical operations, such as roll up, drill down, slice, dice, pivot, and more, which are very applicable for performance analysis, drift detection, and performance tracing. The OLAP systems can be easily integrated with data visualization tools and alerting systems to provide a comprehensive ML observability solution.

Note   Open source OLAP systems

Open source OLAP systems, such as Apache Pinot, Apache Druid, ClickHouse, and more, are increasingly popular and widely adopted across many big companies, including Uber, Netflix, Cisco, LinkedIn, and more. Their ability to support large, complex analytical workloads with blazing-fast speed makes them ideal for applications like ad hoc analysis, real-time analytics, BI reporting, and business metrics dashboards.

Nearly all these OLAP systems support real-time data ingestion, which powers a wide range of real-time data-driven use cases. These include fraud detection and monitoring, business metrics monitoring, performance monitoring, and travel data analysis. This capability opens doors to a vast array of use cases across various industries.

Case Studies

One interesting observation in the ML observability area is that the commercial vendor solutions from ML observability–focused companies have matured significantly over the last few years to provide and their solutions can easily integrate with either in-house or cloud vendor ML infrastructure. Additionally, numerous open source solutions have emerged in recent years to help with various needs of ML observability.

The following sections will highlight a few case studies from both in-house and open source communities.

Lyft: Model Monitoring

In early 2020, the ML platform team at Lyft developed a robust ML monitoring system to identify and prevent model degradation. This system was aimed to minimize negative impact on Lyft's riders, drivers, and financial performance.

The Lyft ML platform team, in their blog,[5] outlined the model monitoring approaches they developed and the culture shift needed for ML practitioners to effectively monitor their models. These areas offer valuable learning lessons and practical insights that go beyond the typical ML observability vendor documentation. As such, the following sections will focus on those two crucial aspects, providing a deeper understanding of how to implement and maintain effective model monitoring practices.

Recognizing that more complex, high-value capabilities take longer to implement, the implementation of their model monitoring system followed a two-phase strategy. First, they prioritized monitoring techniques that are fast to onboard models and to catch the most obvious model problems. Then, they shifted focus to building more powerful offline monitoring techniques for deeper analysis and proactive problem identification.

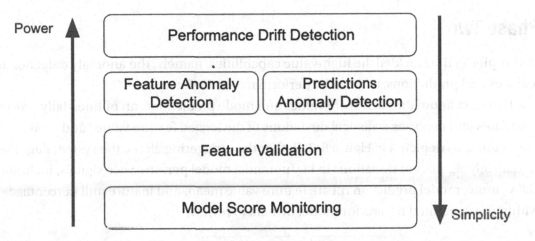

Figure 6-5. *Adapted from Spectrum of Model Monitoring Techniques[5]*

Phase One

In this phase, they focused on model score monitoring and feature validation. For model score monitoring, their system provides an easy and configurable time series–based alert on the average model prediction score that falls within a certain value range. With the flexibility of this approach, each model owner can easily set the appropriate alert

---

[5] Mihir Mathur and Jonas Timmermann, Full-Spectrum ML Model Monitoring at Lyft, 2022, https://eng.lyft.com/full-spectrum-ml-model-monitoring-at-lyft-a4cdaf828e8f

threshold for their models. For feature validation, their system provides a set of out of the box commonly used feature validations, such as type check, value range, missing values, set membership, and required features. Advanced and custom feature validations are supported through custom logic provided by model owners. The combined set feature validation logic is executed against the features of every prediction request. This feature validation is based on the expectation concept, and they leverage the Great Expectations open source library, which will be covered in the next section.

While Great Expectations validators excel at handling large datasets, their overhead hindered their effectiveness in online predictions scenarios where single-feature set validation with sub-milliseconds response time is crucial. To address this, they built a lightweight validation version, leveraging asynchronous execution to minimize latency impact on model prediction speed.

Phase Two

In this phase, they tackled the high-value capabilities, namely, the anomaly detection of features and predictions, and model performance drift detection.

To detect anomalies of feature values and mode predictions, an offline, daily process calculates and analyzes statistical deviations of the logged feature values and model predictions, as depicted in Figure 6-6. Rather than triggering alerts, the system flags the statistically significant deviations in key potential model performance signals, including call volume, model prediction mean, feature value mean, and feature null percentage, within the generated reports for model owners to review.

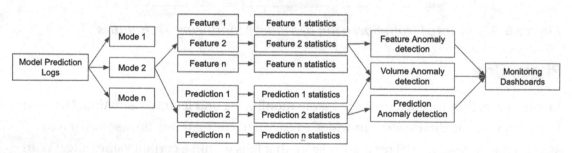

Figure 6-6. *Adapted from Anomaly Detection System Components[5]*

Their automatic statistical check approach requires no onboarding from model owners.

Similar to the anomaly detection approach, their performance drift detection approach is also done on an offline and scheduled basis. The drift detection is done by comparing the model predictions with the ground truth. Since the ground truth is different for each model, the model owners are required to provide the following three pieces of information:

- An SQL query for retrieving the ground truth

- Post-processing logic for computing the performance metrics

- A collection of validations for those metrics

This drift detection is quite effective at detecting complex model performance issues. However, it is a high-touch process and applicable to use cases where the ground truth is available and reliable.

Cultural Shift for Adoption

One of the learnings their blog calls out is about the necessary culture changes needed for driving adoption of their ML monitoring approaches.

Unless it is baked into the model development process, operational concerns of models in production are typically not top of mind for model owners. To drive the adoption of their ML monitoring offerings, the ML platform team at Lyft employed a multi-pronged approach:

- They invested significant effort in making onboarding as smooth as possible.

- They invested time in selling and educating the monitoring offerings to their internal customers through brown bags and partnering with product teams.

- Once there was a healthy organic adoption, they make monitoring for all models mandatory going forward.

Similar to other organizations, they recognize building an ML monitoring is a defensive investment, and it can be challenging to prioritize this investment over other initiatives that have a clearer top-line or bottom-line impact. Oftentimes, it requires a noticeable production incident with financial consequences to reveal the true value of a monitoring system.

Open Source

Numerous open source projects related to ML observability have emerged in recent years. The three open source projects that are highlighted in this section cover the wide spectrum of ML observability infrastructure needs. They range from data-related validations for data and feature generation to logging and capturing key statistical properties of the predictions, and finally to evaluate, test, and monitor ML models from validation to production.

Great Expectations

High quality data is the lifeblood of machine learning. Testing and validations of data and features are crucial to many steps in the ML development lifecycle, including data ingestion, feature engineering, model inference, and more. Figure 6-7 highlights the steps that need data validation and testing. This ensures models make accurate predictions and organizations get the ROI from their successful ML projects.

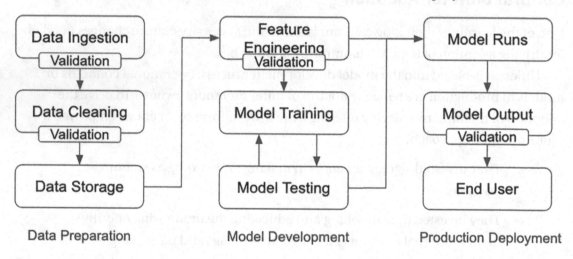

Figure 6-7.   *Adapted from How does Great Expectations fit into ML Ops?*[6]

Great Expectations is a Python open source project designed as a data validation and documentation framework. Its creators believe that data quality issues shouldn't hinder innovation, and their philosophy revolves around solving these issues through proactive data validation and documentation. In other words, they advocate for building data pipelines with data quality "baked in," not "bolted on" later.

---

[6] How does Great Expectations fit into ML Ops?, 2020, `https://greatexpectations.io/blog/ml-ops-great-expectations`

The core concept in the Great Expectations framework is called "Expectation," and it is designed to allow users to clearly and simply define what they expect from their data in a declarative way, making it easy to read and understand at first glance. In essence, each Expectation declares an expected state of the data in a human-readable way, and is meaningful for both technical and non-technical users.

Note   User feedback

For some ML applications, user feedback is one form of validation that can be used to improve model performance. The above steps are meant to highlight the common validation and testing steps involved in bringing models to production.

The Great Expectations framework, offered as a Python library, contains a vast array of built-in Expectation types, ranging from simple table row counts and missingness of values to complex distribution of values, means, and standard deviations. Designed with flexibility and extensibility in mind, the framework enables its user community to easily create custom Expectation types when unique needs arrive. To explore the complete and up-to-date list of Expectations, please visit `https://greatexpectations.io/expectations/`.

The code snippet, Listing 6-1, illustrates a few examples of using table-level and column-level Expectations.

Listing 6-1. Examples of table-level and column-level expectations

```
# import and setting up data to perform validation
import great_expectations as gx
validator = gx.read_csv("<data file location>")
# table-level Expectations
validator.expect_table_row_count_to_be_between(min_value=100, max_
value=300)

# column-level Expectations
validator.expect_column_values_not_to_be_null("salary")
validator.expect_column_values_not_to_be_unique("id")
validator.expect_column_max_to_be_between("age", min_value="100", max_
value="120")
validator.expect_column_mean_to_be_between("age", 20,40)

# validator.get_expectation_suite()
```

There are many more useful and powerful Expectations for the various data validation needs.

In the context of MLOps pipelines, the Great Expectations Python library can be easily integrated into these pipelines to perform various data validations at different steps in the ML development lifecycle. The results of validations can be translated into human-readable reports that can be easily reviewed offline or shared with various data stakeholders. An example of such a report is depicted in Figure 6-8 for two simple and successful validations for column passenger_count and pickup_datetime.

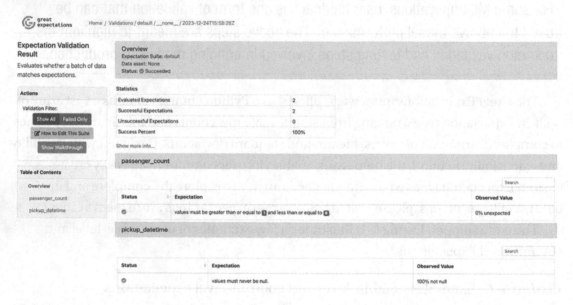

Figure 6-8. *An example of the validation report in human readable format*

whylogs

Studies have shown that high-quality data fuels the model's performance. However, real-world data constantly changes, potentially leading to model performance degradation if the model is not continuously monitored and updated. The ability to easily and quickly detect and monitor shifts in the underlying data distribution is crucial to maintaining model performance in production.

whylogs is an open source library used to generate statistical summaries of various data types, including tabular data, text, audio, images, and more. Instead of employing data sampling techniques, this library uses the data profiling techniques to accurately represent rare events and outliers even in the presence of a large volume of logs. The resulting statistical summaries, known as whylogs profiles, are captured in a standard format for representing a snapshot of data over intervals of time. They provide a comprehensive overview of key statistics like value distributions, distinct value counts, quantiles, frequent items, and more. Additionally, whylogs profiles allow for customization with user-defined metrics and metadata.

The whylogs profiles make it easy to monitor data quality, track changes in the dataset over time, and visualize. These three usages enable a variety of ML observability use cases, including feature drift or concept drift detection, performance degradation analysis, model inputs or data pipeline quality validation, and more.

whylogs profiles are designed with a clear separation between the process of producing profiles and the process of acting upon them, enabling flexibility and scalability in data analysis or alerting. The profiles are designed to be efficient, customizable, and mergeable, making them suitable for a variety of scenarios with strict requirements, such as feature and ML pipelines.

- Efficient: whylog profiles are very compact in size and efficiently describe the datasets they represent. All statistics are collected in a streaming fashion that requires only a single pass over the data with minimal memory overhead. This enables whylog profiles to linearly scale with the number of columns instead of with the data volume.

- Customizable: whylogs profiles offer easily configurable and customizable statistics, such as custom trackers for different data types and use cases. This flexibility is crucial for accommodating diverse metrics and enables the tracking of complex data types such as text and images.

- Mergeable: whylogs profiles can be generated in a distributed manner and then combined together at a later step to form new profiles which represent the aggregate of their constituent profiles.

Note   Data sampling vs. data profiling

In the world of microservices, logging has been adopted as a standard practice for understanding and monitoring of the health of a complex system. In the data domain, data logging is designed to achieve a similar purpose of understanding and monitoring of data. There are two approaches to data logging: sampling and profiling.

Sampling is a simpler approach between the two, and it is done by randomly or programmatically selecting samples of data from a larger data stream and storing them for later analysis. This simple approach is straightforward to implement; however, there are drawbacks when it comes to capturing important statistical properties of the overall dataset, such as missing rare events and outliers and inaccurately estimating the min/max values or the number of unique values.

Profiling approach is about data profiling[7] (also referred to as data sketching or statistical fingerprinting) to capture a human interpretable statistical profile of a given dataset to provide insight into the data. The profiling is done by leveraging a family of efficient streaming algorithms to generate scalable, lightweight statistical profiles of datasets. The captured data profiles could accurately represent rare events and outliers, and statistical properties such as histogram, mean, and standard deviation, are easily interpretable.

The code snippet in Listing 6-2 shows how easy it is to generate whylogs profiles in Python, assuming the whylogs Python package was already installed.

Listing 6-2. An example of using whylogs Python APIs to generate whylogs profiles

```
# import and setting up data to perform validation
import whylogs as why
import pandas as pd
df = pd.read_csv("<data file location>")
# generate profiles
data_profile = why.log(df)
# inspect profiles in a Pandas Dataframe format
data_profile_view = data_profile.view()
data_profile_df = data_profile_view.to_pandas()
```

---

[7] Isaac Buckus, Sampling isn't enough, profile your ML data instead, 2020, https://towardsdatascience.com/sampling-isnt-enough-profile-your-ml-data-instead-6a28fcfb2bd4

By default, the number of rows in the generated whylogs profiles will be equal to the number of columns in the logged dataset. The standard generated metrics of each column in the logged dataset are stored in columns of each row of the generated whylogs profiles. The standard metrics fall into these buckets: counts, cardinality, distribution, types. For more details about the generated metrics, please see this documentation page at https://whylogs.readthedocs.io/en/latest/examples/basic/Inspecting_Profiles.html.

To visualize the summary and drift report of the generated whylog profiles, we will use the provided NotebookProfileVisualizer class. See Listing 6-3 for an example of using this class to display these reports for the Titanic dataset.

Listing 6-3. An example of using whylogs Python APIs to generate a drift report

```
# import and setting up data to perform validation
import whylogs as why
import pandas as pd
titanic_df = pd.read_csv("titanic.csv")
# separate the titanic_df into two using the Survived value for the purpose
of generating a drift summary
# titanic profile
titanic_profile = why.log(titantic_df)
profile = titanic_profile.profile()
prof_view = profile.view()
# reference whylogs profile
cond_reference = (titanic_df['Survived']==0)
titanic_reference = titanic_df.loc[cond_reference]
# drop the Survived and Name columns
titanic_reference = titanic_reference.drop(["Survived","Name"],  axis=1)
ref_result = why.log(pandas=titanic_reference)
ref_prof_view = ref_result.view()
# target whylogs profile
cond_target = (titanic_df['Survived']==1)
titanic_target = titanic_df.loc[cond_target]
# drop the Survived and Name columns
titanic_target = titanic_target.drop(["Survived","Name"],  axis=1)
target_result = why.log(pandas=titanic_target)
```

```
target_prof_view = target_result.view()
# instantiate NotebookProfileVisualizer to generate report
from whylogs.viz import NotebookProfileVisualizer
visualization = NotebookProfileVisualizer()
visualization.set_profiles(target_profile_view=target_prof_view, reference_
profile_view=ref_prof_view)
# generate drift report
visualization.summary_drift_report()
```

Figure 6-9 shows a generated drift report from two whylogs profiles. This is meant as an example to illustrate the distribution shift by visually inspecting the distributions.

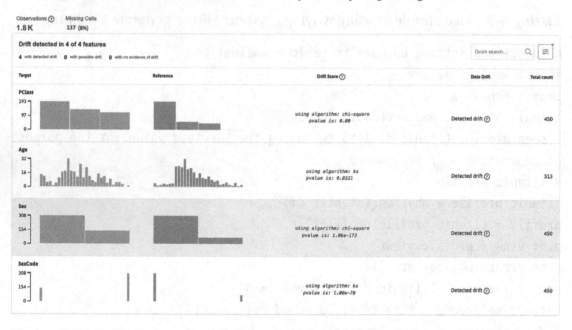

Figure 6-9. *Drift report example from the two whylogs profiles generated from Titanic dataset*

To facilitate a closer visualization of the distributions of individual features, we can overlay two distributions using histogram chart for numerical features or distribution chart for categorical features. Listing 6-4 shows code snippets for how to do that, and Figure 6-10 shows the two charts.

Listing 6-4. Using whylogs Python APIs to overlay two distribution in histogram and distribution charts

```
# continue from Listing 6-3
# Overlay two distributions for numerical feature "Age" in a double
histogram chart
visualization.double_histogram(feature_name="Age")
# plot distribution chart for categorical feature "Sex"
visualization.distribution_chart(feature_name="Sex")
# generate feature statistics of feature called "Age"
visualization2 = NotebookProfileVisualizer()
visualization2.set_profiles(target_profile_view=prof_view, reference_
profile_view=None)
visualization2.feature_statistics(feature_name="Age")
```

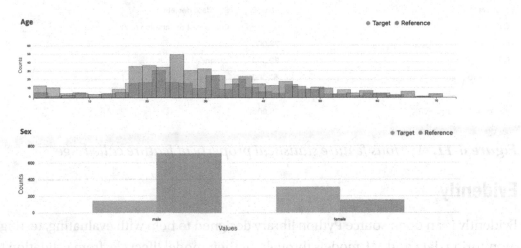

Figure 6-10. *Histogram and distribution charts for feature Age and Sex, respectively*

The preceding two charts visually indicate the distributions of feature Age and Sex in the reference and target whylogs profiles are very different.

It is quite easy to generate the statistics of one or more features. Listing 6-5 shows how to do that with just a few lines of code, and the result is shown in Figure 6-11.

Listing 6-5. Using whylogs Python APIs to generate feature statistics

```
# continue from Listing 6-4
# generate feature statistics for feature called "Age"
visualization2 = NotebookProfileVisualizer()
visualization2.set_profiles(target_profile_view=prof_view, reference_
profile_view=None)
visualization2.feature_statistics(feature_name="Age")
```

Age: Summary Statistics

Distinct (%)	Missing	Mean	Minimum	Maximum
9.92	**557**	**30.398**	**0.170**	**71.000**

Quantile statistics		Descriptive statistics	
5-th percentile	6.000	Standard deviation	14.259
Q1	21.000	Coefficient of variation (CV)	0.47
median	28.000	Sum	22980.88
Q3	39.000	Variance	203.32
95-th percentile	57.000		
Range	70.830		
Interquartile range (IQR)	18.000		

Figure 6-11. *Various feature statistical property of feature called Age*

Evidently

Evidently is an open source Python library designed to help with evaluating, testing, and monitoring data and ML models throughout their model lifecycle, from validation to production. It contains the following concepts: Metric, Reports, Test, and Test Suites.

Metrics are a central component in Evidently, designed to evaluate specific aspects of the data or model quality, such as the number data missing values, statistical properties, and more. There are numerous built-in metrics related to data quality, integrity, drift, and model performance.

A Report is a combination of different metrics that evaluate data or ML model quality. The report output can be visualized with interactive graphs or can be generated in JSON format or Python dictionary. Reports are commonly used for debugging and exploratory purposes and ad hoc analysis through their interactive visualizations.

In Evidently, a Test is basically a metric with a condition, used to verify expectations about the data or model, whether that is about the structure of the data or the model performance. Multiple Tests can be organized into a Test Suite and are evaluated together. Tests and Test Suites are suitable for automating the expectations about data quality and model performance to ensure they are not violated.

The core components in Evidently are Metric, Report, Test, and Test Suites, together addressing a wide range of usage scenarios, from ad hoc analysis and automated pipeline testing to continuous monitoring.

Out of the box, Evidently Python library provides many commonly needed Metrics and Tests. Known as Presets, these ready-to-use components streamline data and model evaluation and testing for common scenarios. A comprehensive list of available Presets, along with detailed descriptions, can be found at this web page `https://docs.evidentlyai.com/presets/all-presets`.

Listing 6-6 shows code snippets for using Evidently Python APIs to generate a data drift report and the statistical properties of column Age from the Titanic dataset.

Listing 6-6. Using Evidently Python APIs to generate and visualize a drift report

```
# set up the necessary import statements
import pandas as pd
from evidently.report import Report
from evidently.metric_preset import DataDriftPreset
from evidently.metrics import DatasetSummaryMetric, ColumnSummaryMetric,
ColumnQuantileMetric
# read the Titanic dataset
titantic_df = pd.read_csv("../data/Titanic.csv")
# drop the column "Unnamed: 0"
titantic_df = titantic_df.drop("Unnamed: 0", axis=1)
# separate by target value
cond_reference = (titantic_df['Survived']==0)
titanic_reference = titantic_df.loc[cond_reference]

cond_target = (titantic_df['Survived']==1)
titanic_target = titantic_df.loc[cond_target]
# drop the two unnecessary columns
titanic_reference = titanic_reference.drop(["Survived","Name"], axis=1)
titanic_target = titanic_target.drop(["Survived","Name"], axis=1)
```

```
# set up and run the DataDriftPreset report
report = Report(metrics=[
    DataDriftPreset(),
])
report.run(reference_data=titanic_reference, current_data=titanic_target)
report.show(mode='inline')

# generate data summary metrics for column Age
report2 = Report(metrics=[
    DatasetSummaryMetric(),
    ColumnSummaryMetric(column_name='Age'),
    ColumnQuantileMetric(column_name='Age', quantile=0.25),
])

report2.run(current_data=titantic_df, reference_data=None)
report2.show(mode='inline')
```

Figure 6-12 contains the drift report of the columns: Ages, PClass, SexCode, and Sex, similar to the ones in Figure 6-9. Figure 6-13 shows the generated statistical properties of a feature called "Age."

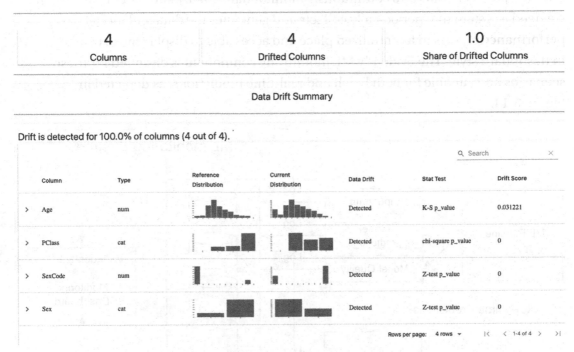

Figure 6-12. *The generated drift report for Age, PClass, SexCode, and Sex features*

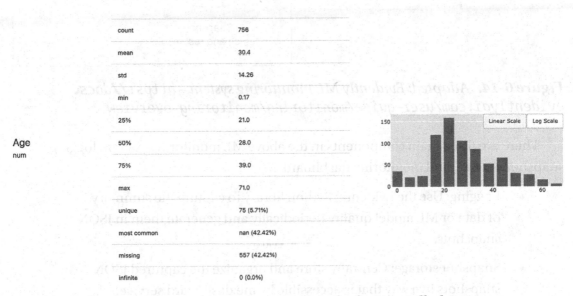

Figure 6-13. *The various statistical properties of a feature called Age*

The Reports and Test Suites outputs can be reviewed and interactively visualized once they are generated. To continuously monitor data quality and model performance, we need a system that periodically collects and generates the data and model performance reports in a centralized place and accessible to display those results in dashboard style. The Evidently ML monitoring component is designed for these scenarios and suitable for both batch and real-time predictions, as depicted in Figure 6-14.

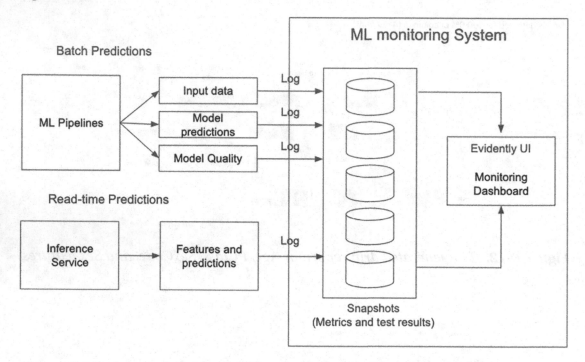

Figure 6-14. *Adapted: Evidently ML monitoring system –* `https://docs.`
`evidentlyai.com/user-guide/monitoring/monitoring_overview`

There are three main components in the above ML monitoring system: logging, snapshot storage, and monitoring dashboard.

- Logging: Use the Evidently Python library to capture the summary of data or ML model quality periodically and generate them in JSON snapshots.

- Snapshot storage: Centrally store and organize the captured JSON snapshots in a way that is accessible by the dashboard service.

- Monitoring dashboard: Use Evidently Monitoring service to parse the snapshots and visualize the captured metrics or test results in the interactive graphs.

Evidently ML monitoring component provides a basic blueprint for organizations seeking to leverage ML observability open source solutions to incrementally build their ML observability infrastructure. This component is a work in progress and evolving. For most up-to-date details and progress, please visit their documentation page at `https://docs.evidentlyai.com/user-guide/monitoring/monitoring_overview`.

One key and common aspect of the three highlighted open source solutions in this section is that they are classified as vendor-backed open source. Each is supported by a commercial company, offering benefits like roadmap, stability, and customer support, and often accompanied by a cloud-based commercial version with additional features.

Summary

ML observability plays a key role in ensuring that the ML model performance remains at the expected level after they are deployed to production. Model performance degradation can occur gradually over time for various reasons, such as data drift or model drift. These degradations can have a significant negative business impact, affecting aspects ranging from monetary value to user experience and company reputation.

ML observability infrastructure is a crucial pillar of ML infrastructure. It is designed to detect, monitor, alert, and diagnose model performance issues. A central component of this infrastructure is the observability store. It is meant to capture a diverse type of model-related data that are produced across the stages of the ML development process, including feature engineering, model training, model evaluation, and model prediction. This assortment of data is what makes the observability store unique and valuable and to meet the needs of ML observability: quickly surfacing up model performance issues and reducing time to resolving those issues. Other components that an ML observability infrastructure needs are data and metric ingestion and computation, alert and notification, and interactive visualization. The pyramid of monitoring AI/ML solutions,[8] as depicted in Figure 6-15, illustrates the need for many types of monitoring to ultimately support ML observability.

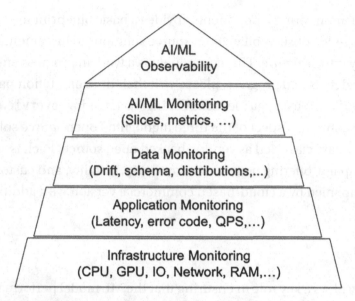

Figure 6-15. *The pyramid of monitoring for AI/ML solutions*

The main parts that ML observability needs to support include feature validation, model prediction monitoring, feature and prediction anomaly detection, and data and model drift detection. Open source solutions such as Great Expectations, whylogs, and Evidently are available for organizations to leverage and incorporate into their ML observability infrastructure to address some of these key aspects.

Three years ago, organizations had no choice but to build their own ML observability infrastructure from the ground up. Today, there are around two dozen companies offering their vendor solutions.[9] These solutions have matured, become more sophisticated, and somewhat commoditized. The integration with these vendor solutions has become increasingly straightforward. Unless an organization has very specific ML observability needs or is operating at a scale that these vendor solutions can't support, it is sensible to include a vendor solution evaluation as a part of the build vs. buy strategy.

---

[8] Mederic Hurier, Is AI/ML Monitoring just Data Engineering, 2023, `https://mlops.community/is-ai-ml-monitoring-just-data-engineering-%F0%9F%A4%94/`

[9] Ruth Sherdan, ML Observability - Hype or Here to Stay?, 2022, `https://medium.com/at-the-front-line/ml-observability-hype-or-here-to-stay-acef064ff843`

## CHAPTER 7

Ray Core

As discussed in previous chapters, MLOps is becoming essential in today's AI landscape. A central challenge in MLOps is to manage ML-related **computation**, including loading data from external storage, performing necessary last-mile transformations, training ML models, tuning hyperparameters, and generating inference results.

Many companies that begin their machine learning journey often start with a single machine to perform these computation tasks. For example, a single `p4d.24xlarge` instance on AWS EC2 has 8 NVIDIA A100 GPUs with 40 GB GPU memory per GPU – an enormous number crunching machine!

However, the approach of using "a single beefy GPU machine" has a few severe limitations, and almost all companies will eventually need distributed computing in MLOps.

The first challenge is **data**. Deep Learning algorithms are "hungry" for data in training because as the volume and complexity of their data grow, the limitations of a single machine become apparent. At the time when this chapter was written, the most powerful GPU chip was the NVIDIA A100, with either 40Gb or 80Gb of GPU memory (GRAM), and 400 Gbps network bandwidth. Once the volume of data reaches TB level, it quickly becomes infeasible to load and process the data on a single machine.

The second factor is **price** (and availability of the largest GPU instances). Again, let's use NVIDIA GPUs as data points. On EC2 (the same conclusion holds for Azure and GCP), it is disproportionately more expensive to rent a single very powerful GPU (like A100), compared with renting several less powerful GPUs (like A10G). What's more, it is very difficult to reserve the most powerful GPUs even if someone is willing to pay the high price, as illustrated in Figure 7-1.

247

© Hien Luu, Max Pumperla and Zhe Zhang 2024
H. Luu et al., *MLOps with Ray*, https://doi.org/10.1007/979-8-8688-0376-5_7

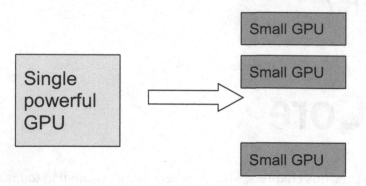

Figure 7-1. *Ray allows using a collection of small GPU nodes instead of a single powerful GPU*

And here enters Ray (`https://docs.ray.io/`) to provide **easy access to flexible distributed computing on heterogeneous hardware**. Ray is a distributed computing framework to support scalable execution of computer programs on a cluster. Ray was created at UC Berkeley in 2016 as an open source software (OSS) project to support advanced machine learning projects (including real-time machine learning). The project has been one of the fastest-growing open source distributed computing frameworks in the past few years.

This chapter first introduces "Ray Core," which is the general-purpose framework to scale out Python programs.[1] In the next chapter, we will introduce our suite of Ray AI Libraries (they each support a specialized workload, e.g., model training). Figure 7-2 illustrates different Ray components, including Ray Core and Ray AI libraries.

---

[1] Ray Core also supports Java and C++, but for the purposes of this chapter, we will focus on Python, which is the lingua franca in ML and MLOps.

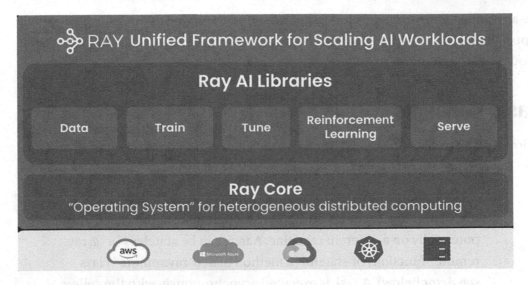

Figure 7-2. *Ray Core and Ray libraries*

Ray Core in a Nutshell

Basically, Ray Core provides **asynchronous/concurrent program execution on a cluster scale**.

Most developers are familiar with some form of asynchronous (or concurrent) programming. Examples include Java `Future` and Python `multiprocessing`. These primitives are fundamental to modern computing – allowing programs to perform multiple tasks simultaneously and efficiently use the available resources. From web servers handling multiple requests at the same time to multi-core processors executing parallel code, concurrency is essential for building high-performance applications that can handle complex workloads.

However, existing concurrent programming paradigms are **limited to a single computer**. The central motivation of the Ray project is to take the concept and power of concurrent programming to the next level, and allow computation to span multiple machines and heterogeneous computing devices.

Ray Core aims to be a general-purpose distributing computing framework. As a result, it is designed to handle all distributed computing concerns such as communication, fault tolerance, resource allocation, and so forth (so that these concerns are abstracted away from developers). It is beyond the scope of this book to cover all aspects of Ray Core. The rest of this chapter will cover the most relevant information about Ray in the context of MLOps.

Basic Concepts

Below are the basic application concepts a developer should understand in order to develop and use Ray programs:

- Task: A remote function invocation. This is a single function invocation that executes on a process different from the caller, and potentially on a different machine. A task can be stateless (a "@ray. remote" function) or stateful (a method of a "@ray.remote" class – see *Actor* below). A task is executed asynchronously with the caller: the ".remote()" call immediately returns one or more "ObjectRefs" (futures) that can be used to retrieve the return value(s).

- Object: An application value. These are values that are returned by a Task/Actor, or created through "ray.put". Objects are **immutable**: they cannot be modified once created. A worker can refer to an object using an "*ObjectRef*."

- Actor: A stateful worker process (an instance of a "@ray.remote" class). Actor tasks must be submitted with a *handle*, or a Python reference to a specific instance of an actor, and can modify the actor's internal state during execution.

- Driver: The program root, or the "main" program. This is the code that runs "ray.init()".

- Job: The collection of tasks, objects, and actors originating (recursively) from the same driver, and their runtime environment. There is a 1:1 mapping between drivers and jobs.

API Basics

In this section, we first provide the most basic introduction to Ray Core APIs and show how to use them in Python code. Table 7-1 contains a conceptual introduction of the most basic Ray APIs. The rest of this section will provide more details.

Table 7-1. *Basic Ray APIs (in Python)*

API	Functionality
`ray.init`	Initializes the "Ray context" and makes sure the current process is connected to a Ray Cluster. See SparkContext for reference
`@ray.remote`	Decorates Python functions and classes so they become Ray Tasks and Actors to be executed on a cluster
`.remote`	Postfix to *trigger* the *asynchronous* execution of remote function calls and class instantiations
`ray.put(x)`	Puts a Python object x in the Ray Object Store and makes sure the object can be accessed by other Ray Tasks and Actors
`ray.get(y)`	Blocking call that waits for the Python object to be *resolved*
	If y involves a function call, `ray.get` will block until the execution of the function itself, and all other required function calls to finish
	Instead, if y was earlier stored in Ray object store ray.get will block until the Python object is retrived from the Ray object store.

Below are some examples to illustrate the basic APIs. We will first show an example of scaling Python functions – a function is the basic unit of *stateless* computation (generating an output with an input, without maintaining and updating any internal state that could alter the output).

In the below example, the first part is vanilla Python code that runs on a single thread. It basically defines a function that counts the number of lines in a file. Then the main part of the program calls the function on two separate files, and calculates the total number of lines in these two files. The second part illustrates how to parallelize this program with Ray.

```
# Vanilla Python code without Ray
def count_lines(file):
    // count the number of lines in the file as n
    return n
a = count_lines(file_1)
b = count_lines(file_2)
c = a + b

# Using Ray to parallelize the above code
import ray
@ray.remote
def count_lines(file):
    // count the number of lines in the file as n
    return n

a = count_lines.remote(file_1)
b = count_lines.remote(file_2)
c = ray.get(a) + ray.get(b)
```

Figure 7-3 illustrates how Ray implements asynchronous computing.

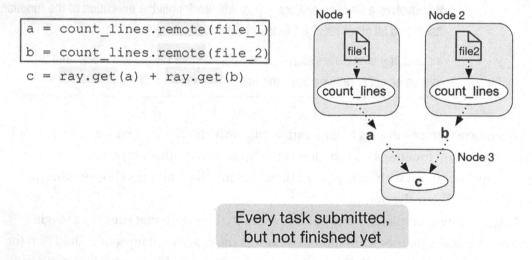

Figure 7-3. *Ray implements asynchronous computing*

Basically, when a = count_lines.remote(file_1) is called, the actual compute tasks are *submitted* (subject to Ray Scheduling). Completion of these compute tasks is only guaranteed when c = ray.get(a) + ray.get(b) is called (especially the ray.get part) – again, this is similar to the get call in Java Future. This is illustrated in Figure 7-4.

```
a = count_lines.remote(file_1)
b = count_lines.remote(file_2)
c = ray.get(a) + ray.get(b)
```

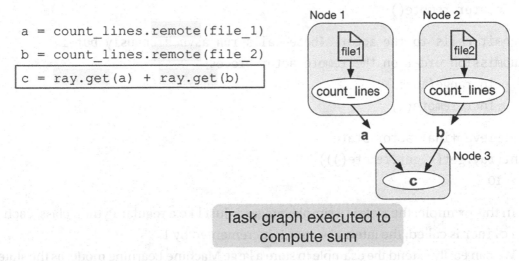

Task graph executed to compute sum

Figure 7-4. Task graph executed

The above example is simple enough. But in Machine Learning (and therefore in MLOps), it is often important to have *stateful* computations as well – for example, maintaining an *attribute* in a Python class that can be initialized and updated, and subsequently used in generating outputs in member functions.

For instance, when doing model inference, we want to download the model once and use it to handle every piece of input data. When doing model training, it's oftentimes beneficial for each trainer to retain information across multiple training iterations or batches – the model's current parameters, gradients, or other relevant variables.

Below is a simple example of using Ray Actors to perform stateful computation.

```
import ray

@ray.remote
class Counter:
    def __init__(self):
        self.i = 0

    def get(self):
        return self.i
```

```
    def incr(self, value):
        self.i += value

# Create a Counter actor.
c = Counter.remote()

# Submit calls to the actor. These calls run asynchronously but in
# submission order on the remote actor process.
for _ in range(10):
    c.incr.remote(1)

# Retrieve final actor state.
print(ray.get(c.get.remote()))
# -> 10
```

In this example, the *state* is a simple integer i. Just like a regular Python class, each time c.incr is called, the internal state i is incremented by 1.

We can easily extend the example to store a large Machine Learning model as the state. These kinds of pretrained large models are becoming very popular and prevalent in the current era of LLMs and generative AI. The following example shows how to host three replicas of the GPT-2 model (in three Ray Actors) concurrently and generate responses from these replicas in parallel (potentially leveraging multiple servers/VMs with GPU chips).

```
import ray
import random

# Define an Actor (that uses one CPU and one GPU) to generate responses
from prompts, based on the GPT-2 model
@ray.remote(num_cpus=1, num_gpus=1)
class Model:
    def __init__(self):
        # Download the GPT-2 model which is > 700MB
        from transformers import pipeline
        self.model = pipeline("text-generation", model="gpt2")
    def respond(self, prompt: str) -> str:
        return(self.model(prompt, max_length=20, num_return_sequences=5))

# Create 3 replicas of the model
models = [Model.remote() for _ in range(3)]
```

```
# A long list of prompts
prompts = ["Tell me a joke", "Tell me a lullaby", "Best restaurant in San
Francisco", "Hello", "Best movie", "Peter Pan's best friend"]

# Submit these prompts to the 3 replicas (the 3 replicas will perform the
GPT-2 computation in parallel
futures = []

for p in prompts:
    i = random.randint(0, 2)
    futures.append(models[i].respond.remote(p))

# Finally, wait until all responses are generated
print(ray.get(futures))
```

Architecture Basics

As a cluster computing framework, Ray's architecture has many similarities with earlier frameworks such as Apache Hadoop, Apache Spark, Kubernetes, and so forth. Interested readers are encouraged to go through Ray's architecture white paper https://bit.ly/ray-arch. In this section, we will cover the most relevant aspects in the context of using Ray in MLOps. Figure 7-5 is a simplified diagram of the main components of a Ray cluster.

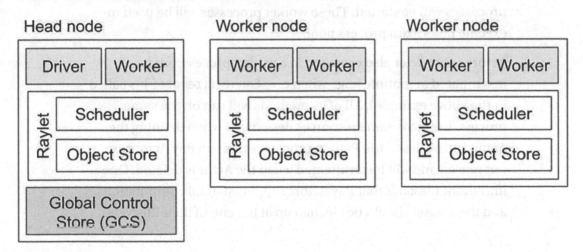

Figure 7-5. A Ray Cluster

Fundamentally, when a **Ray cluster is started** on a number of "nodes" (e.g., physical servers, cloud virtual machines, Kubernetes containers). A raylet process is started on each node. Each raylet is a long running "daemon" process whose lifetime is usually the same as the cluster itself. It is similar to the NodeManager concept in Apache Hadoop, or how "sidecars" are used in the Kubernetes ecosystem. The main responsibilities of a raylet are threefold:

- Managing **worker processes**, including scheduling Tasks to worker processes (see below).

- Managing resources on the node, including memory and dependencies (e.g., which Python packages should this node install).

- In particular, it manages part of the memory in the physical computer. Collectively, the memory managed by all raylets in the cluster forms the Ray Object Store, which is a cluster-wide in-memory store to provide a global namespace for all Ray Objects.

Each Ray node also has a number of **worker processes**. Each worker process is a Python process that is used to execute Tasks or run dedicated Ray Actors:

- Tasks: When a Ray cluster is started, by default, on each node, one worker process is started for each CPU. For instance, for a Ray cluster of 8 nodes, where each node has 32 CPUs, a total of 256 worker processes will be started. These worker processes will be used to execute tasks like a process pool.

- Actor: A Ray Actor also executed in a worker process, but is instantiated at runtime (e.g., when c = Counter.remote() is called in the above example). All of its methods will run on the same process, using the same resources designated when defining the Actor. Unlike Tasks, the Python processes that run Ray Actors are not reused and will be terminated when the Actor is deleted. One important reason is that Ray Actors support stateful computations, and their states should be cleaned up at the end of their lifecycles.

For readers who are familiar with Apache Spark, a *node* in a Ray cluster is similar to a "worker node" in a Spark cluster. Within a node, a Ray **worker process** is similar to a Spark executor for the part of executing **Tasks**. But a Ray **worker process** can also execute a stateful **Actor**, which is not supported in Apache Spark.

Among all the Ray nodes, one node is designated as the **head node**. The most important responsibility of the **head node** is managing the **Global Control Service** (GCS). The **head node/GCS** is responsible for managing certain important global metadata (for instance, location of Actors). More details can be found on `https://bit.ly/ray-arch`.

Then, with a running Ray cluster, a developer can send code (usually Python) to run on it. The starting point of any Ray application is the *Driver* process (which is launched on one of the Ray nodes).

The responsibilities of the Driver process directly map to the basic APIs that Ray provides (please refer to the API Basic section). The Python code in the Driver process can create one of the following:

- Task: When a Python *function* is decorated with `@ray.remote` and invoked with `remote`, it becomes a *Task*, which is the basic unit of stateless computation in Ray.

- Actor: When a Python *class* is decorated with `@ray.remote` and initiated with `remote`, it becomes an *Actor*, which is the basic unit of stateful computation in Ray (with the attributes of the class representing the Actor state).

- Object: In Ray, an object refers to either of the following:

  - The return of a Ray Task/function call in an Actor

  - References (that can be used later) as a result of `ray.put()`

  All Ray objects are stored in the Ray Object Store.

Fundamentally, Ray's architecture is designed to support flexible and efficient asynchronous computing. Two main concepts are Ownership and Dependency. We use the following figure to illustrate how these two concepts are implemented in Ray, via the translation of Python code ➤ logical task graph ➤ physical execution graph.

Ownership is the concept using which Ray manages metadata in a decentralized way. This concept means that each **Object** (see above definition for Objects) in the application will be managed by a single **worker process**. This worker (as the "owner") is responsible for ensuring execution of the task that creates the value and facilitating the resolution of the reference of an **Object** to its underlying value. In the following example, the **worker process** that runs **Task** a owns the **Task** b (because b is started by a). Therefore, that **worker process** is responsible for making sure **Task** b is completed and the return value is obtained.

Dependency is a simpler concept. In asynchronous computation, it means that the completion of one **Task** *depends* on the completion or resolution of several other **Tasks** or **Objects**. In the following example, **Task** a cannot complete before the Object x is resolved (the value obtained by a). This is illustrated in Figure 7-6.

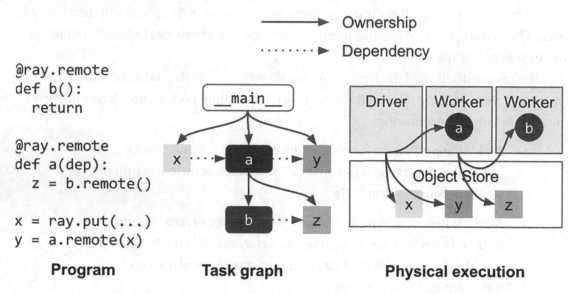

Figure 7-6. *Ownership and dependency in Ray*

Scheduling

Scheduling is a critical aspect of any distributed system, and this is particularly true for Ray, since Ray aims to become the unified framework for building distributed applications. In particular, Ray's scheduling mechanisms are designed for compute- and data-intensive workloads. In this section, we will explore the inner workings of the Ray scheduler and how it enables efficient task scheduling in distributed environments. We will examine the different scheduling strategies that Ray employs, and how they can be customized to meet the specific needs of different applications.

To understand how scheduling works in Ray, and the implications in MLOps, let's first look at the lifetime of a **Task** and an **Actor**.

A Task's lifetime starts when either a process (either the Driver or another Task/Actor) calls a Python function decorated with @ray.remote. The process which makes the call becomes the *Owner* of the task. It is the Owner's responsibility to negotiate with Raylets to acquire necessary resources for the Task. For example, if a Task is decorated with @ray.remote(num_cpus=2), the Owner needs to send this resource request to one or more Raylets, and finally find a **worker process** that has at least 2 CPUs available. At this point, this **worker process** is considered *leased* to this Owner – and the Raylet on the same node as the worker process remembers this lease information. Then the Owner directly works with the leased worker process to transfer necessary information (e.g., arguments of the function) to get the Task started. This is illustrated in Figure 7-7.

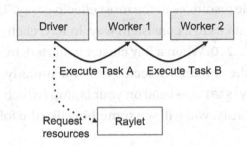

Figure 7-7. *A tasks's lifetime*

The lifetime of an Actor is pretty different from the lifetime of a Task – and the biggest difference is the involvement of the Global Control Service (GCS). In the current Ray architecture, GCS maintains the metadata of each Actor. Therefore, the creation of each Actor always goes through the GCS. When, for example, the Driver process calls c = Counter.remote(), the Driver first *registers* the Actor with the GCS. Then, the GCS – in contrast with the Owner process when creating a Task – is responsible for requesting resources from Raylets, and actually creating the Actor. This is illustrated in Figure 7-8.

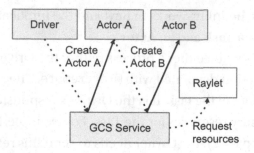

Figure 7-8. *An Actor's lifetime*

With the preceding background information, now let's understand Ray's scheduling policies (which are very important when running ML workloads).

In essence, scheduling is about matching a Task/Actor's resource requirements with a Ray node's available resources in the most effective way. The first basic step is to represent and define resources in a consistent way. In Ray, each *resource* is a key-value pair – for instance, "GPU": 2.0. When a Ray cluster is started, by default, the *resources* on each Ray node are set to the amounts detected from the underlying operating system.[2] For example, if you do `ray start --head` on your laptop (which starts a single-node Ray cluster), and then `ray status`, you will see something like the following:

```
Resources
-------------------------------------
Usage:
 0.0/10.0 CPU
 0B/13.20GiB memory
 0B/2.00GiB object_store_memory
```

Subsequently, when Tasks/Actors are created, their resource requirements come from the decorator – for instance, `@ray.remote(num_cpus=2)`. An important note is that resource requests are *logical* in Ray. In the current implementation, physical resource limits are not enforced by Ray.

In finding the best location of a Task/Actor, in addition to making sure the resource requirements can be fulfilled by the destination node, there are a few other important considerations. The first consideration is data locality. Ray is often used for

---

[2] This default can be overridden by customized configurations.

data-intensive workloads such as data pre-processing before model training. Therefore, it is important that Tasks are created on the same node where the required data is stored (e.g., arguments of the Task).

Consider the following example, where two Python functions (Ray Tasks) are defined and executed. The first Task read_array_from_file reads an input file and returns an ndarray. Like we discussed in the Architecture section, this return value is stored in the Ray Object Store. The second Task double_array takes the returned ndarray as input, and generates a new ndarray with each value doubled. When the double_array Task is scheduled, Ray tries to place it on the same node that has the return value of read_array_from_file physically stored.

```
import ray

@ray.remote
def read_array_from_file(file):
    // read the file into an ndarray
    return arr

@ray.remote
def double_array(arr):
    return arr * 2

arr = read_array_from_file.remote(file1)
doubled_arr = ray.get(double_array.remote(arr))
print(doubled_arr)
```

Another scheduling consideration is the balance between *Packing* and *Spreading*. This is a classic topic in cluster computing in general, even beyond Ray. On the high level, Packing refers to fulfilling the resource requirements of submitted Tasks/Actors on *as few nodes as possible*. The main benefit is to increase resource utilization and to allow downsizing the cluster and therefore reducing the cost. *Spreading* refers to placing the Tasks/Actors on all available nodes in a round-robin fashion – this helps load balancing.

Lastly, an advanced option in Ray scheduling is *Placement Groups*. A Placement Group allows a Ray user to reserve a group of resources across multiple nodes in a transactional manner. This is highly useful in certain Machine Learning scenarios where progress cannot be made without a minimum set of resources (known as the "gang scheduling" scenario). We won't go into more details about this advanced scheme here, and interested readers should refer to `https://bit.ly/ray-arch`.

Fault Tolerance

Ray is a sophisticated distributed system, and that means the occurrence of failures is a norm rather than exception. Generally, there can be two types of failures: (1) Ray-level failures and (2) Application-level failures. Let's understand these two types of failures from the perspective of software abstractions or contracts.

Now that the readers are familiar with the basic concepts in Ray's APIs and architecture, we can look at the `count_lines` example again to understand these two types of failures. As illustrated in the following figure (in a simplified manner), if the Raylet on Node 1 crashes (e.g., killed by the OS for unexpected reasons), this would be a Ray-level failure. In contrast, if file1 is too large for the Python function `count_lines` and causes an OutOfMemory (OOM) error, this would be an Application-level failure.

Here's another intuitive way to understand this. Ray is responsible for *scaling out* Python functions and classes to a cluster of multiple computers; therefore, any error on the scaling out part is regarded as a Ray-level failure. Any error happening within the Python function or class[3] (that Ray tries to scale out) is regarded as an application-level failure. Figure 7-9 is a simple diagram to demonstrate this framework.

---

[3] The same error would occur anyways even if you run the same Python function or class locally on that computer.

```
a = count_lines.remote(file_1)
b = count_lines.remote(file_2)
c = ray.get(a) + ray.get(b)
```

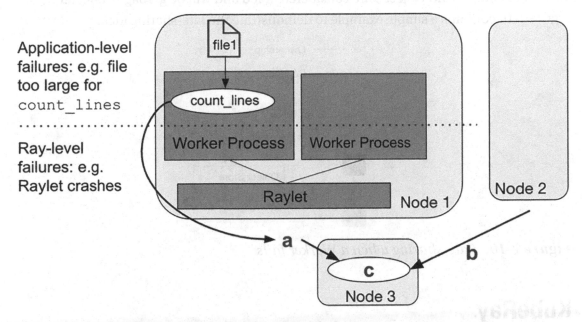

Figure 7-9. *Ray-level and Application-level failures*

For Ray-level failures: Ray's general approach is to rely on retries and reconstructions to automatically recover from these failures. Ray provides fault tolerance for Tasks, Actors, and Objects in similar ways with varying details, as explained in Table 7-2.

Table 7-2. *The fault tolerance policies for Ray Tasks, Actors, and Objects*

	Fault Tolerance Policy
Task	By default, a failed Ray Task will be retried for 3 times. This can be customized with `@ray.remote (max_retries=x)`
Actor	By default, a failed Ray Actor will not be restarted. But this can be customized with `@ray.remote (max_retries=y)`
Object	Ray will try to reconstruct a lost object by re-executing necessary Tasks. This is called "lineage-based reconstruction"

For Application-level failures, the best practice is to catch and handle error codes from the application code (this is a responsibility of the application developer).

Another important fault tolerance concept to understand is *fate-sharing*. A Ray Task/Actor/Object is always considered failed (and will be garbage-collected) once the Owner process fails. In the following example, if the Worker process running Task a fails, Task b and Object z are considered failed and will be garbage-collected. Figure 7-10 contains a simple example to demonstrate the fate-sharing idea.

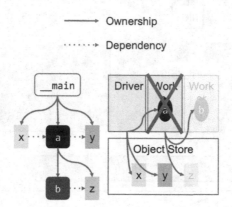

Figure 7-10. *Fate-sharing when a Worker fails*

KubeRay

Kubernetes has become the standard deployment framework for MLOps due to its ability to manage and orchestrate complex containerized workloads in a distributed environment. As machine learning models and pipelines have become more sophisticated, their deployment has become increasingly complex, requiring the integration of multiple tools (sometimes in different programming languages such as Python and Java). Kubernetes provides a unified platform for managing containerized applications, with automated scaling, load balancing, and fault tolerance.

The KubeRay project (https://ray-project.github.io/kuberay/) is the standard way of deploying and managing Ray clusters on Kubernetes.

Following the basic principles of Kubernetes, KubeRay basically packages key Ray components into "pods." Please see the Architecture section for details.

A central component of KubeRay is the "KubeRay Operator." This Operator is responsible for starting and maintaining the lifetime of other Ray pods – headnode pod, worker node pods, and the autoscaler pod (responsible for increasing or decreasing the size of the cluster). In particular, for online serving/service scenarios (which is becoming more popular now), the KubeRay operator is responsible for making sure the Ray Headnode pod is highly available. Figure 7-11 is a simple illustration of KubeRay's architecture.

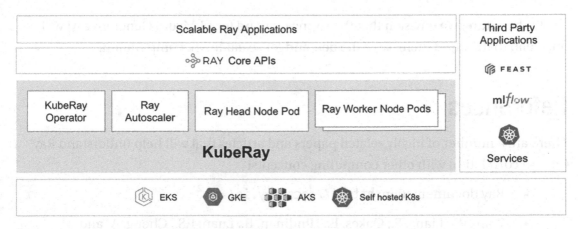

Figure 7-11. *Architecture of KubeRay*

Summary

In summary, in today's MLOps ecosystem (dominated by Python), Ray Core is a leading framework to scale the computation across multiple CPU/GPU servers (and other hardware accelerators).

Without using Ray, ML and MLOps developers are not able to fulfill two conflicting requirements: (1) iterate quickly on ML models and (2) each iteration requires setting up the compute infrastructure with a large amount of resources (CPU, GPU, memory). Ray Core gives developers the "superpower" to iterate quickly at scale, through architectural innovations that are explained in this chapter (e.g., scheduling Python functions and classes in a cluster). In order to provide this simplicity to developers, Ray Core handles the complexities of distributed computing, including fault tolerance, resource management, scheduling, and so forth.

In the recent two to three years, Ray has been widely adopted by leading MLOps teams in the industry, including Uber, DoorDash, and Netflix.

In the future, we envision that the strong demand for LLM and Generative AI will only lead to the rapid increase of the demand for scalable ML computation.

References

There are a number of highly related papers and articles that will help understand Ray Core in correlation with other computing concepts:

- Ray documentation: `https://docs.ray.io/`

- Wang, S., Liang, E., Oakes, E., Hindman, B., Luan, F.S., Cheng, A. and Stoica, I., 2021, April. Ownership: A Distributed Futures System for Fine-Grained Tasks. In *NSDI* (pp. 671-686).

- Wang, S., Hindman, B. and Stoica, I., 2021, June. In reference to RPC: it's time to add distributed memory. In *HotOS* (pp. 191-198).

- Ray v2 Architecture (`https://bit.ly/ray-arch`)

CHAPTER 8

An Introduction to the Ray AI Libraries

Overview

In the last chapter you've learned about Ray Core, and how it can be used to build distributed applications. This chapter introduces you to the main concepts of Ray's Artificial Intelligence (AI) libraries and how you can utilize them for creating and deploying typical ML workflows. We will demonstrate how to use those libraries by building an application that fine-tunes a large language model (LLM), deploys it for online inference, and uses it for offline batch inference. We will also explain when and why you should use Ray's AI libraries and provide a brief overview of the Ray AI ecosystem. Lastly, we will delve into the connection between Ray and other systems.

What Are Ray's AI Libraries?

Ray's AI libraries are a suite of tools that can be effectively used together in common ML workflows. You can see them as a comprehensive toolkit designed to support your ML workloads by providing various third-party integrations for tasks like model training and accessing custom data sources. These libraries abstract away lower-level complexities and offer an intuitive API. The following libraries are available:

- Ray Data for data ingest and processing: To use machine learning effectively, you must prepare the data in a way that the ML model can understand. This process involves selecting and transforming the data that will be input into the model. It can be a challenging task, so it is helpful to have access to reliable tools to assist with it. Ray

© Hien Luu, Max Pumperla and Zhe Zhang 2024
H. Luu et al., *MLOps with Ray*, https://doi.org/10.1007/979-8-8688-0376-5_8

Data helps you load, transform, and consume your data in a scalable manner. This library is also your choice for tasks such as batch prediction of data on a trained model.

- Ray Train and Ray RLlib for model training: For machine learning, it is necessary to train your algorithms on data that has been previously processed. This involves selecting the appropriate algorithm for the task at hand. Having a diverse range of algorithms to choose from can also be beneficial. For supervised training, you can use Ray Train, and if you want to run reinforcement learning (RL) experiments, RLlib is your natural choice.

- Ray Tune for hyperparameter tuning: During the process of training a machine learning model, certain parameters can be fine-tuned in order to improve its performance. In addition to these model parameters, there are also hyperparameters that can be adjusted before training begins. The proper adjustment of these hyperparameters can significantly impact the effectiveness of the final machine learning model. Tools like Ray Tune help you with the process of optimizing these hyperparameters.

- Ray Serve for model serving: The deployment of trained models is necessary in order to provide access to them. Serving a model involves making it available through various means, such as using HTTP servers. Ray Serve was built for scalable deployment of your ML models.

This list is not exhaustive, and there is more to consider when building machine learning applications, but these four steps (loading and processing data, model training, hyperparameter tuning, and model serving) are critical for the success of your ML projects. As indicated, Ray has solutions for every step of this process, and it's important to stress that all these libraries are distributed by design, as part of the Ray ecosystem.

Importantly, the four steps above are usually not run in isolation, but rather as part of a common data science process. Clearly, it is beneficial for all components to be working smoothly together. Ray's AI libraries were designed with this in mind, providing a common runtime and API for experiments and the capability to expand workloads as needed. That means all discussed libraries are sufficiently interoperable and can be used both for prototyping and production, in which topics such as scalability and reliability matter. See Figure 8-1 for a concise summary of all AI libraries and the use cases they cover.

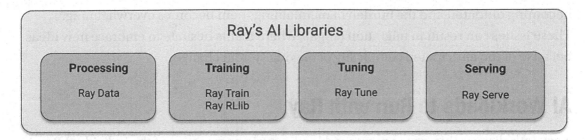

Figure 8-1. Ray's AI libraries, designed to tackle your ML use cases at scale

Why and When to Use Ray for ML?

Over the past few years, running ML workloads with Ray has undergone continuous development. Initially, Ray RLlib for RL workloads and Ray Tune for hyperparameter optimization (HPO) were created on top of Ray Core. Subsequently, additional components like Ray Train for ML model training, Ray Serve for serving trained models, and more recently, Ray Data for data processing were introduced. The creation of these libraries was the outcome of active discussions and feedback from the ML community.

The Ray AI libraries cater to both data scientists and ML engineers. Data scientists can utilize them to build and scale their end-to-end experiments, as well as handle individual tasks such as preprocessing, training, tuning, scoring, or serving ML models. ML engineers have the flexibility to build a customized ML platform with Ray AI libraries or simply leverage them individually to integrate with other libraries in their ecosystem. Additionally, Ray allows users to drop down and work directly with the lower-level Ray Core API for greater flexibility. Ray, being a Python-native tool with strong GPU support and stateful primitives (Ray actors) suitable for complex ML workloads, is a natural choice for providing higher-level libraries for ML workloads.

Ray allows you to smoothly move from experimenting on a laptop to running production workflows on a cluster. Typically, data science teams hand over their ML code to production teams, which can be costly and time-consuming due to the need for code modifications or rewrites. Ray simplifies this transition by handling scalability, reliability, and robustness concerns for you.

By choosing Ray's AI libraries, you can avoid the challenge of dealing with multiple distributed systems and the complicated glue code that comes with it. When working with numerous components, teams often face the problem of integrations quickly becoming outdated and the burden of maintaining them becomes overwhelming. These issues can result in migration fatigue, where teams hesitate to embrace new ideas because of the anticipated complexity of making system changes.

AI Workloads to Run with Ray

Next, let's discuss this from a more workload-centric perspective of the individual Ray AI libraries we already discussed. Ray is specifically designed to handle common tasks in AI projects, which can be classified as follows:

- Stateless computation: Tasks like preprocessing data or making model predictions on a batch of data are considered stateless. These workloads can be computed independently in parallel, and Ray tasks are used to handle them. Many big data processing tools fall into this category.

- Stateful computation: On the other hand, tasks like model training and hyperparameter tuning involve updating the model state during the training process. Ray actors are used to handle the complexities of distributing stateful workers for such operations.

- Composite workloads: AI workloads often involve a combination of stateless and stateful computation. For example, processing features and then training a model. Ray is built to efficiently handle these advanced composite workloads, known as big data training, by effectively managing both the stateless and stateful aspects.

- Online serving: Ray also excels at scalable online serving of one or more models. The transition from the previous three types of workloads to serving is seamless within the Ray ecosystem.

These types of workloads can be used in various scenarios. For example, you can use Ray to replace and scale a single component of an existing pipeline. Additionally, you have the option to create your own end-to-end machine learning applications using Ray. Moreover, as we will discuss later, you can even build your own AI platform with the help of Ray's AI libraries. Figure 8-2 briefly summarizes the AI workloads Ray was built for.

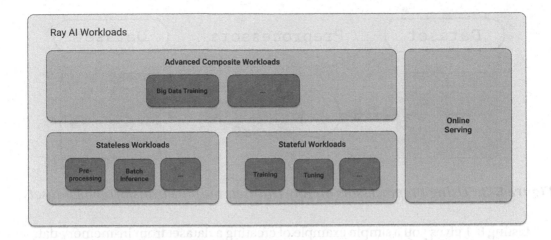

Figure 8-2. *Ray is built to handle a wide range of AI workloads*

An Introduction to Ray's AI Libraries

Ray's design philosophy for its AI libraries is to enable you to handle complex machine learning workloads, ideally just using a single Python script and a single distributed system in Ray. Before diving into a longer example that shows you all libraries in action, let's first take a brief look at the core abstractions in isolation, starting with data ingestion and processing.

Datasets and Preprocessors

In Ray, the standard way to load and process data is by using so-called Ray Datasets from the Ray Data library. To prepare input data for machine learning experiments, implementations of the `Preprocessor` class are used. These preprocessors transform your data into features. By operating on Datasets and leveraging the Ray ecosystem, these preprocessors enable efficient scaling of preprocessing steps.

During training, you "fit" a Preprocessor to the specified training data, which can then be subsequently used for both training and serving purposes. Ray Data includes numerous built-in preprocessors that cover a wide range of use cases. However, if you require a specific preprocessor that is not available, it is straightforward to define a custom preprocessor according to your needs. Figure 8-3 schematically shows how Preprocessors act on Datasets to output transformed datasets.

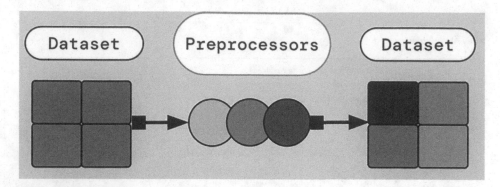

Figure 8-3. *Using Preprocessors on Ray Datasets outputs transformed Datasets*

Listing 8-1 gives you a simple example of creating a dataset from in-memory data, and scaling it to a range of [0, 1] by using `fit_transform` on a dedicated Preprocessor called MinMaxScaler. We will discuss loading of larger datasets and more realistic processing later on. Note that the code in Listing 8-1, and all other code snippets in this chapter assume that you've installed Ray version 2.7.0, for instance, with `pip install "ray[data,train,tune,serve]==2.7.0"`.

Listing 8-1. Creating a Ray Data Dataset in Python and scaling it with a Preprocessor

```
import ray
from ray.data.preprocessors import MinMaxScaler

ds = ray.data.range(10)
preprocessor = MinMaxScaler(["id"])

ds_transformed = preprocessor.fit_transform(ds)
print(ds_transformed.take())
```

The following table summarizes the most common preprocessors available in Ray Data, together with their type (Table 8-1). Some of the preprocessors listed there we'll see again in a later section.

Table 8-1. *Common Ray Data preprocessors and their types*

Preprocessor Type	Ray Data Preprocessors
Feature scalers	MaxAbsScaler, MinMaxScaler, Normalizer, PowerTransformer, RobustScaler, StandardScaler
Generic preprocessors	Concatenator, Preprocessor, SimpleImputer
Categorical encoders	Categorizer, LabelEncoder, OneHotEncoder, MultiHotEncoder, OrdinalEndcoder

Trainers

Once you have your Ray Datasets defined, you can proceed to specify a Trainer with Ray Train. A Trainer is responsible for running a machine learning algorithm on your data. It acts as a consistent interface for training frameworks like TensorFlow, PyTorch, or HuggingFace. In the example in the next section, we will focus on HuggingFace Transformers, but it's important to note that other framework integrations work the same way using the Ray Train API.

In essence, Ray Trainers offer scalable ML training on Datasets. Additionally, they are designed to integrate well with Ray Tune for hyperparameter optimization, which we will discuss in the next section. Figure 8-4 shows you how a Trainer works in principle. Besides providing Datasets, you also want to provide a `ScalingConfig` before calling `fit` on your Trainer, which will create the distributed training process for you.

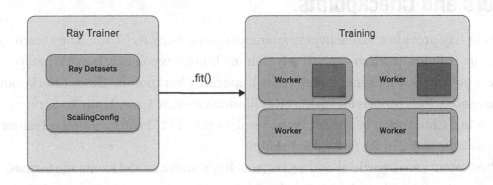

Figure 8-4. *Instantiating and calling Trainers to run distributed training on your data*

In the following code example, we use Ray's XGBoost integration and define an XGBoostTrainer to train a simple model on synthetic, tabular data. For this specific type of trainer, we also need to define a label_column. We also make sure to specify the learning objective, by passing a dictionary to the params argument to the XGBoostTrainer constructor in Listing 8-2. Note that it is mandatory for the datasets argument to have a "train" key, but you could additionally pass in keys like "validation", "test", or custom datasets as well.

Listing 8-2. Running XGBoost model training with Ray Train

```
import ray
from ray.train.xgboost import XGBoostTrainer
from ray.train import ScalingConfig

train_dataset = ray.data.from_items(
    [{"x": x, "y": 2 * x} for x in range(0, 32, 3)]
)

trainer = XGBoostTrainer(
    label_column="y",
    params={"objective": "reg:squarederror"},
    scaling_config=ScalingConfig(num_workers=2),
    datasets={"train": train_dataset},
)

result = trainer.fit()
```

Tuners and Checkpoints

Tuners in Ray provide scalable hyperparameter tuning using Ray Tune. They seamlessly work with Trainers and also support any training functions you may have. When you run Trainers or Tuners, they generate framework-specific Checkpoints. These Checkpoints can be used to load models for various Ray libraries such as Tune, Train, or Serve. You can obtain a Checkpoint by accessing the result of the .fit() call on either a Trainer or a Tuner, as we will see in the next code example.

Checkpoints are beneficial as they serve as Ray's native model exchange format. They allow you to easily retrieve trained models at a later stage without worrying about custom methods to store and load the models. Figure 8-5 illustrates how Tuners interact with Trainers to produce Checkpoints.

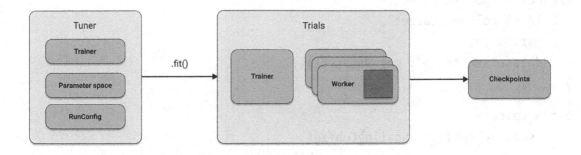

Figure 8-5. *Ray Tuners tune the hyperparameters of Ray Trainers on a set of workers, and they also generate Checkpoints for you*

To illustrate how running a Tuner works in Python, for sake of simplicity let's use an XGBoostTrainer again, but this time use a slightly more involved setup in params, and use a larger dataset to operate on. Apart from the Trainer instance, our Tuner in Listing 8-3 takes a `param_space` argument to inform the Tuner which hyperparameters to tune, and a `run_config` that tells the Tuner how to tune the training experiment.

Listing 8-3. Running hyperparameter optimization using a Tuner on an XGBoost model

```python
from sklearn.datasets import load_breast_cancer

from ray import tune
from ray.data import from_pandas
from ray.train import RunConfig, ScalingConfig
from ray.train.xgboost import XGBoostTrainer
from ray.tune.tuner import Tuner

def get_dataset():
    data_raw = load_breast_cancer(as_frame=True)
    dataset_df = data_raw["data"]
    dataset_df["target"] = data_raw["target"]
    dataset = from_pandas(dataset_df)
    return dataset
```

```
trainer = XGBoostTrainer(
    label_column="target",
    params={},
    datasets={"train": get_dataset()},
)

param_space = {
    "scaling_config": ScalingConfig(
        num_workers=tune.grid_search([2, 4]),
        resources_per_worker={
            "CPU": tune.grid_search([1, 2]),
        },
    ),
    "params": {
        "objective": "binary:logistic",
        "tree_method": "approx",
        "eval_metric": ["logloss", "error"],
        "eta": tune.loguniform(1e-4, 1e-1),
        "subsample": tune.uniform(0.5, 1.0),
        "max_depth": tune.randint(1, 9),
    },
}
tuner = Tuner(trainable=trainer, param_space=param_space,
    run_config=RunConfig(name="my_tune_run"))

# Run tuning job and get a checkpoint
result = tuner.fit()
checkpoint = result.checkpoint
```

Running Batch Prediction

Batch inference refers to generating model predictions on a set of input data. The model is usually a complex ML model, like a neural network, but could just be a simple Python function. In batch inference, also known as offline inference, your model is run on a large batch of data on demand. This is in contrast to online inference, where the model is run immediately on a data point when it becomes available.

Running batch inference in Ray is conceptually easy and requires three steps. First, you load your data and apply any preprocessing you need. Then you define your model and a transformation that applies your model to your data. And lastly, you simply use map_batches from Ray Data to apply the transformation on your data. Figure 8-6 illustrates this process schematically.

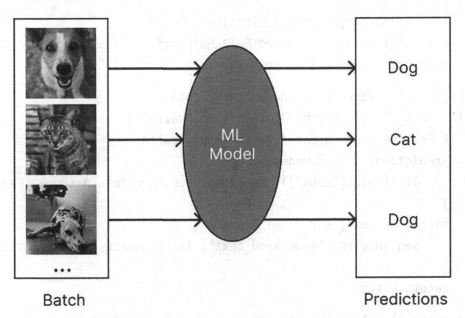

Batch   Predictions

Figure 8-6. *In batch inference, batches of data are mapped against a model to get predictions on demand*

Let's quickly illustrate this process by loading a simple GPT-2 model from HuggingFace to define a predictor class with it in Listing 8-4. We create synthetic text data and load it into a Ray Dataset using the from_numpy helper from Ray Data. Finally, we use map_batches to get predictions. Note how we employ a so-called ActorPoolStrategy to scale out this batch inference task to as many resources as we like.

Listing 8-4. Running a simple batch prediction job on a HuggingFace transformer model

```
from typing import Dict
import numpy as np

import ray
```

```
# Step 1: Create a Ray Dataset from in-memory Numpy arrays.
ds = ray.data.from_numpy(np.asarray(["Complete this", "for me"]))

# Step 2: Define a Predictor class for inference.
class HuggingFacePredictor:
    def __init__(self):
        from transformers import pipeline
        # Initialize a pre-trained GPT2 Huggingface pipeline.
        self.model = pipeline("text-generation", model="gpt2")

    # Logic for inference on one batch of data.
    def __call__(self, batch: Dict[str, np.ndarray]):
        # Get the predictions from the input batch.
        predictions = self.model(
            list(batch["data"]), max_length=20, num_return_sequences=1
        )
        batch["output"] = [
            sequences[0]["generated_text"] for sequences in predictions
        ]
        return batch

# Use 2 parallel actors for inference. Each actor predicts on a
# different partition of data.
scale = ray.data.ActorPoolStrategy(size=2)

# Step 3: Map the Predictor over the Dataset to get predictions.
predictions = ds.map_batches(HuggingFacePredictor, compute=scale)

# Step 4: Show one prediction output.
predictions.show(limit=1)
```

Online Serving Deployments

Instead of using batch prediction "offline" and interacting with the model in question directly, you can leverage Ray Serve to deploy an inference service that you can query over HTTP/S. The core abstraction to serve models with Ray Serve is the Deployment.

To see a Deployment in action, let's work with a simple model from HuggingFace that translates English to French in Listing 8-5, and use the @serve.deployment decorator from Ray Serve to make it a Deployment.

Listing 8-5. Running a Ray Serve application with a HuggingFace translation model

```
# File name: serve_quickstart.py
from starlette.requests import Request

import ray
from ray import serve

from transformers import pipeline

@serve.deployment(num_replicas=2, ray_actor_options={"num_cpus": 0.2, "num_
gpus": 0})
class Translator:
    def __init__(self):
        # Load model
        self.model = pipeline("translation_en_to_fr", model="t5-small")

    def translate(self, text: str) -> str:
        # Run inference
        model_output = self.model(text)

        # Post-process output to return only the translation text
        translation = model_output[0]["translation_text"]

        return translation

    async def __call__(self, http_request: Request) -> str:
        english_text: str = await http_request.json()
        return self.translate(english_text)

translator_app = Translator.bind()
```

You can run this example by calling the command serve run serve_
quickstart:translator_app, which will host the model on http://127.0.0.1:8000/.
To test the model, simply execute the Python script in Listing 8-6 in a separate terminal window.

Listing 8-6. Getting a translation from our served translation model

```
import requests

english_text = "Hello world!"

response = requests.post("http://127.0.0.1:8000/", json=english_text)
french_text = response.text

print(french_text)
```

This completes our whirlwind tour of Ray AI libraries from a bird's eye view. Next, let's delve into a more sophisticated example to show you how Data, Train, Tune, and Serve, can work together to tackle a highly relevant use case.

An Example of Training and Deploying Large Language Models with Ray

In this extended example, we will showcase how to use Ray for fine-tuning a language model called GPT-J-6B. GPT-J is a language model trained on the Pile dataset[1] and is conceptually derived from OpenAI's GPT-2 model. There are variants of this model, but this particular one gets its name from having 6 billion parameters. For more information on GPT-J, see the HuggingFace documentation.[2] We will use Ray with its HuggingFace Transformers integration together with a pre-trained model from HuggingFace Hub, but can easily adapt this example to use similar models.

Note that this example is not meant to be run locally on your laptop. Instead, we're trying to convey a realistic scenario for fine-tuning and deploying a medium-sized language model and give you some practical cost and runtime considerations along the way. It should be mentioned that, if you have a good workstation with a modern GPU, appropriately modifying the scaling configurations in this example will allow you to run this example directly, but it would likely take much longer than what we indicate here.

---

[1] Pile DataSet, https://huggingface.co/datasets/EleutherAI/pile

[2] GPT-j HuggingFace documentation, https://huggingface.co/docs/transformers/model_doc/gptj

To give you an overview, we will develop this example following these concrete steps:

- We discuss how to start a Ray cluster that has all the Python libraries it needs installed.

- We load a relatively small dataset to fine-tune our GPT-J-6B model on it.

- We define custom preprocessors needed to transform this dataset.

- We scale out fine-tuning of the model on a sizable compute cluster. This part of the example is based on this Ray example,[3] which you can refer to for more in-depth information.

- Then we check if we get useful predictions from the trained model.

- We then run offline batch processing on a test set. You can learn more about this aspect on the Ray documentation.[4]

- Finally, we deploy the model to get predictions from an endpoint, which you can also learn more about reading this example.[5]

We predict that workflows like this will become more and more common as companies try to develop LLMs specialized on their use cases and data availability, so it's worth understanding what goes into a process like this. Discussing this example in depth will also give us a chance to discuss Ray and all of its libraries in a more holistic way than before.

Starting a Ray Cluster and Managing Dependencies

First off, for this example to run, you have to make sure that all dependencies shown in Listing 8-7 are installed on all nodes of your Ray cluster.

---

[3] GPT-J-6B Fine-Tuning with Ray and DeepSpeed, https://docs.ray.io/en/releases-2.7.0/train/examples/deepspeed/gptj_deepspeed_fine_tuning.html

[4] GPT-J-6B Fine-Tuning with Ray and DeepSpeed, https://docs.ray.io/en/releases-2.7.0/train/examples/deepspeed/gptj_deepspeed_fine_tuning.html

[5] GPT-J-6B Serving with Ray, https://docs.ray.io/en/releases-2.7.0/ray-air/examples/gptj_serving.html

Listing 8-7. Installing Ray and other dependencies for this chapter

```
pip install "ray[data,train,tune,serve]==2.7.0" "accelerate==0.18.0"
pip install "transformers>=4.26.0" "torch>=1.12.0"
pip install "datasets" "evaluate" "deepspeed==0.8.3"
```

The proper way to propagate dependencies is by defining a so-called runtime environment,[6] but we're going to skip this step and assume the Python packages in Listing 8-7 are available to all your Ray cluster nodes. Starting a local Ray cluster is done with a single command, as explained in the last chapter and shown in Listing 8-8. The recommended way to run this example is to use Anyscale,[7] which provides seamless Ray-as-a-service.

Listing 8-8. Starting a local Ray cluster

```
import ray

ray.init()
```

Loading a Dataset and Preprocessing It

We will fine-tune our GPT-J-6B model on the "`tiny_shakespeare`" dataset,[8] comprising 40,000 lines of Shakespeare from a variety of Shakespeare's plays. Our goal is to make the model, which has been trained on modern-day English, better at generating text in the style of Shakespeare. First, let's load the dataset in Listing 8-9, using the HuggingFace Datasets library.

Listing 8-9. Loading a Shakespeare text corpus using HuggingFace Datasets

```
from datasets import load_dataset

hf_dataset = load_dataset("tiny_shakespeare")
```

We will use the Data library from Ray for distributed preprocessing and data ingestion. We can easily convert the dataset obtained from HuggingFace to Ray Data by using the `from_huggingface` function in Listing 8-10.

---

[6] Runtime Environments, `https://docs.ray.io/en/releases-2.7.0/ray-core/handling-dependencies.html`

[7] AnyScale, `www.anyscale.com/`

[8] Tiny shakespeare dataset, `https://huggingface.co/datasets/tiny_shakespeare`

Listing 8-10. Converting a HuggingFace dataset to a Ray Dataset in one line of code

```
import ray.data

ray_datasets = {
    "train": ray.data.from_huggingface(current_dataset["train"]),
    "validation": ray.data.from_huggingface(current_dataset["validation"])
}
```

Note that we split this dataset into `train` and `validation` splits. Within each split, the data is represented as a single string, so we need to do some preprocessing to make it usable for training. To proceed, let's first define two helper functions in Listing 8-11, namely:

- `split_text`: This will split the single string into separate lines, removing empty lines and character names ending with ':' (e.g., `ROMEO:`).

- `tokenize`: This will tokenize the lines using the HuggingFace Tokenizer associated with the model, ensuring each entry has the same length (512) by padding and truncating appropriately. This is needed for training.

Listing 8-11. Defining two preprocessing helper functions

```
from transformers import AutoTokenizer

def split_text(batch: pd.DataFrame) -> pd.DataFrame:
    text = list(batch["text"])
    flat_text = "".join(text)
    split_text = [
        x.strip()
        for x in flat_text.split("\n")
        if x.strip() and not x.strip()[-1] == ":"
    ]
    return pd.DataFrame(split_text, columns=["text"])

def tokenize(batch: pd.DataFrame) -> dict:
    tokenizer = AutoTokenizer.from_pretrained(model_name, use_fast=False)
```

283

```
tokenizer.pad_token = tokenizer.eos_token
ret = tokenizer(
    list(batch["text"]),
    truncation=True,
    max_length=512,
    padding="max_length",
    return_tensors="np",
)
ret["labels"] = ret["input_ids"].copy()
return dict(ret)
```

Next, we will use the `map_batches` API from Ray Data in Listing 8-12 to apply these two preprocessors, which we define to work on Pandas dataframe data by specifying the corresponding `batch_format` to be pandas.

Listing 8-12. Using map_batches to transform our Datasets

```
processed_datasets = {
    key: ds.map_batches(split_text, batch_format="pandas").map_
    batches(tokenize, batch_format="pandas")
    for key, ds in ray_datasets.items()
}
```

Fine-Tuning a Language Model

We can now configure Ray's `TorchTrainer` to perform distributed fine-tuning of the GPT-J-6B model. In order to do that, we specify a `trainer_init_per_worker` function, which creates a HuggingFace Transformers `Trainer`. This `Trainer` will be distributed by Ray using Distributed Data Parallelism[9] (DDP) from PyTorch Distributed[10] under the hood. This means that each worker will have its own copy of the model, but operate on different data. At the end of each step, all the workers will synchronize model updates.

---

[9] Getting Started with Distributed Data Parallel, `https://pytorch.org/tutorials/intermediate/ddp_tutorial.html`

[10] PyTorch Distributed Overview, `https://pytorch.org/tutorials/beginner/dist_overview.html`

GPT-J-6B is a relatively large model and doesn't fit on a GPU with less than 16 GB GRAM. To deal with that issue, we can use DeepSpeed,[11] a library to optimize the training process and allow us to offload and partition training states, reducing overall GRAM usage. Furthermore, DeepSpeed ZeRO-3[12] allows us to load large models without running out of memory.

HuggingFace Transformers and Ray's integration (`TransformersTrainer`) allow you to easily configure and use DDP and DeepSpeed. All you need to do is specify a `deepspeed` configuration argument in the `TrainingArguments` object.[13]

The following code in Listing 8-13 to define a `Trainer` might look intimidatingly long at first, but if you look closely, you'll see that the way we define `trainer_init_per_worker` is relatively simple:

- We first define a couple of training hyperparameters, such as the `batch_size`.

- We then set up a DeepSpeed configuration, which takes up a lot of space. You can skip this part on your first read-through.

- Next, we set up the `TrainingArguments` that will later be passed into our Trainer.

- Then we load the GPT-J model from HuggingFace, together with a tokenizer for preprocessing.

- Finally, we build and return the `Trainer` object.

Listing 8-13. Defining a HuggingFace Transformers Trainer with a DeepSpeed configuration

```
import evaluate
import torch
from transformers import (
    Trainer,
    TrainingArguments,
```

---

[11] Microsoft DeepSpeed, https://github.com/microsoft/DeepSpeed

[12] DeepSpeed ZeRO-3 Offload, www.deepspeed.ai/2021/03/07/zero3-offload.html

[13] HuggingFace TrainingArguments, https://huggingface.co/docs/transformers/en/main_classes/trainer#transformers.TrainingArguments

```
    GPTJForCausalLM,
    AutoTokenizer,
    default_data_collator,
)
from transformers.utils.logging import disable_progress_bar, enable_
progress_bar

from ray import train
from ray.train.huggingface.transformers import (
    prepare_trainer,
    RayTrainReportCallback
)

def trainer_init_per_worker(
        train_dataset, eval_dataset=None,
        **config):
    os.environ["OMP_NUM_THREADS"] = str(
        train.get_context().get_trial_resources().bundles[-1].get("CPU", 1)
    )
    # Enable tf32 for better performance
    torch.backends.cuda.matmul.allow_tf32 = True

    batch_size = config.get("batch_size", 4)
    epochs = config.get("epochs", 2)
    warmup_steps = config.get("warmup_steps", 0)
    learning_rate = config.get("learning_rate", 0.00002)
    weight_decay = config.get("weight_decay", 0.01)
    steps_per_epoch = config.get("steps_per_epoch")

    # Step 2: Define a deepspeed config
    deepspeed = {
        "fp16": {
            "enabled": "auto",
            "initial_scale_power": 8,
        },
        "bf16": {"enabled": "auto"},
        "optimizer": {
```

```
        "type": "AdamW",
        "params": {
            "lr": "auto",
            "betas": "auto",
            "eps": "auto",
        },
    },
    "zero_optimization": {
        "stage": 3,
        "offload_optimizer": {
            "device": "cpu",
            "pin_memory": True,
        },
        "offload_param": {
            "device": "cpu",
            "pin_memory": True,
        },
        "overlap_comm": True,
        "contiguous_gradients": True,
        "reduce_bucket_size": "auto",
        "stage3_prefetch_bucket_size": "auto",
        "stage3_param_persistence_threshold": "auto",
        "gather_16bit_weights_on_model_save": True,
        "round_robin_gradients": True,
    },
    "gradient_accumulation_steps": "auto",
    "gradient_clipping": "auto",
    "steps_per_print": 10,
    "train_batch_size": "auto",
    "train_micro_batch_size_per_gpu": "auto",
    "wall_clock_breakdown": False,
}

# Step 3: Set up the training arguments
training_args = TrainingArguments(
    "output",
```

```
        logging_steps=1,
        save_strategy="steps",
        save_steps=steps_per_epoch,
        max_steps=steps_per_epoch * epochs,
        learning_rate=learning_rate,
        weight_decay=weight_decay,
        warmup_steps=warmup_steps,
        label_names=["input_ids", "attention_mask"],
        push_to_hub=False,
        report_to="none",
        disable_tqdm=True,  # declutter the output a little
        fp16=True,
        gradient_checkpointing=True,
        deepspeed=deepspeed,
    )
    disable_progress_bar()

    # Step 4: Load model and preprocessor
    tokenizer = AutoTokenizer.from_pretrained(model_name)
    tokenizer.pad_token = tokenizer.eos_token

    print("Loading model")

    model = GPTJForCausalLM.from_pretrained(
        model_name, use_cache=False
    )
    model.resize_token_embeddings(len(tokenizer))

    print("Model loaded")

    enable_progress_bar()

    metric = evaluate.load("accuracy")

    train_ds = train.get_dataset_shard("train")
    eval_ds = train.get_dataset_shard("validation")

    train_ds_iterable = train_ds.iter_torch_batches(
        batch_size=batch_size
```

```
)
eval_ds_iterable = eval_ds.iter_torch_batches(
    batch_size=batch_size
)

def compute_metrics(eval_pred):
    logits, labels = eval_pred
    predictions = np.argmax(logits, axis=-1)
    return metric.compute(
        predictions=predictions, references=labels
    )

# Step 5: Create and train a Trainer
trainer = Trainer(
    model=model,
    args=training_args,
    train_dataset=train_ds_iterable,
    eval_dataset=eval_ds_iterable,
    compute_metrics=compute_metrics,
    tokenizer=tokenizer,
    data_collator=default_data_collator,
)

# Add callback to report checkpoints to Ray Train
trainer.add_callback(RayTrainReportCallback())
trainer = prepare_trainer(trainer)
trainer.train()
```

With our trainer_init_per_worker complete, we can now instantiate the
TorchTrainer from Ray in Listing 8-14. We set the scaling_config, controlling the
amount of workers and resources used, and the datasets we will use for training and
evaluation. We also pass the splitter and tokenizer preprocessors we have defined
earlier as an argument, wrapped in a Chain. These chained preprocessors will be
included with the returned Checkpoint, meaning it will also be applied during inference.
We then call the TorchTrainer.fit method to start training with Ray. Afterward, we
save the Result object to a variable, so we can access metrics and checkpoints. Finally,
we use the returned Result object to access metrics and the Ray Checkpoint associated
with the last iteration.

Listing 8-14. Starting the training process using Ray

```
from ray.train.torch import TorchTrainer
from ray.train import RunConfig, ScalingConfig

batch_size = 16
train_ds_size = processed_datasets["train"].count()
steps_per_epoch = train_ds_size // (batch_size * num_workers)

trainer = TorchTrainer(
    train_loop_per_worker=train_func,
    train_loop_config={
        "epochs": 1,
        "batch_size": batch_size,  # per device
        "steps_per_epoch": steps_per_epoch
    },
    scaling_config=ScalingConfig(
        num_workers=num_workers,
        use_gpu=use_gpu,
        resources_per_worker={"GPU": 1, "CPU": cpus_per_worker},
    ),
    datasets=processed_datasets,
    run_config=RunConfig(storage_path=storage_path),
)

trainer.fit()

checkpoint = results.checkpoint
checkpoint
```

Training Runtime Considerations

As we are using data parallelism, each worker operates on its own shard of the data. The batch size we set in TrainingArguments is the per-device batch size. By changing the number of workers used for training, we can change the effective batch size and thus the time needed for training to complete. The effective batch size is calculated as

```
per device batch size * # workers * # gradient accumulation steps
```

That means we can scale the effective batch size by adding more workers, and thereby reduce the time to complete training.

- While the training speedup is not linear due to communication overheads, in practice it can often be close to linear.

- Our processed Shakespeare dataset has a total of 1348 examples, and we set per-device batch size to 16. That means using 16 g4dn.4xlarge worker nodes on AWS,[14] the effective batch size is 256. This is equivalent to 85 steps per epoch. Including initialization, it takes around 2440 seconds to run one epoch with this setup.

- With a total of 32 g4dn.4xlarge nodes, the effective batch size is 512, which equates to 43 steps per epoch and a runtime of about 1280 seconds per epoch.

- If you run this example locally on a single GPU, the effective batch size will go down accordingly, and training speed will decrease proportionally.

Generate Text from a Prompt

We can use the trained model to generate predictions from our fine-tuned model. In Listing 8-15 we use a HuggingFace Transformers pipeline[15] to generate predictions from the fine-tuned model. We set device_map="auto" so that the model is automatically placed on the right device and set the task to "text-generation".

Listing 8-15. Getting predictions from our Checkpoint

```
from transformers import pipeline, AutoTokenizer, GPTJForCausalLM

model = GPTJForCausalLM.from_pretrained("/local/checkpoint")
tokenizer = AutoTokenizer.from_pretrained("/local/checkpoint")

pipe = pipeline(
    model=model,
```

---

[14] Amazon EC2 G4 Instances, https://aws.amazon.com/ec2/instance-types/g4/
[15] HuggingFace Pipelines class, https://huggingface.co/docs/transformers/en/main_classes/pipelines

```
    tokenizer=tokenizer,
    task="text-generation",
    torch_dtype=torch.float16,
    device_map="auto",
)

# Generate from prompts!
for sentence in pipe(["Romeo and Juliet", "Romeo", "Juliet"], do_
sample=True, min_length=20):
    print(sentence)
```

Running Batch Inference for Our GPT-J Model

Instead of getting predictions on a small sample dataset, let's now discuss how to do batch prediction on a potentially much larger Ray Dataset. With Ray you run three steps to carry out batch inference:

1. Load a Ray Data dataset and apply any preprocessing you need. This will distribute your data across the cluster.

2. Define your model in a class and define a transformation that applies your model to your data batches (of format Dict[str, np.ndarray] by default).

3. Run inference on your data by using the ds.map_batches() method from Ray Data. In this step you also define how your batch processing job gets distributed across your cluster.

Let's start by defining a simple Dataset consisting of text prompts to feed into an LLM in Listing 8-16.

Listing 8-16. Loading a toy dataset into Ray Data for batch prediction

```
import ray.data
import pandas as pd

prompt = (
    "In a shocking finding, scientists discovered a herd of",
    "unicorns living in a remote, previously unexplored valley"
)
```

```
ds = ray.data.from_pandas(
    pd.DataFrame([prompt] * 10, columns=["prompt"])
)
```

For the purposes of this example, we will use a very small toy dataset composed of multiple copies of our prompt. Ray Data can handle much bigger datasets with ease. Additionally, we'll be using the same GPT-J-6B model we used before. The simplest way is to use map_batches with a callable class, as shown in Listing 8-17, which represents a user-defined function (UDF). This will allow us to save time by initializing a model just once and then feed it multiple batches of data. All we need is a class that initializes our model and has a __call__ method to predict on batches of data.

Listing 8-17. Defining a callable class UDF for batch prediction

```
model_id = "EleutherAI/gpt-j-6B"
revision = "float16"  # use float16 weights to fit in 16GB GPUs

class PredictCallable:
    def __init__(self, model_id: str, revision: str = None):
        from transformers import AutoModelForCausalLM, AutoTokenizer
        import torch

        self.model = AutoModelForCausalLM.from_pretrained(
            model_id,
            revision=revision,
            torch_dtype=torch.float16,
            low_cpu_mem_usage=True,
            device_map="auto",  # automatically makes use of all GPUs
            available to the Actor
        )
        self.tokenizer = AutoTokenizer.from_pretrained(model_id)

    def __call__(self, batch: pd.DataFrame) -> pd.DataFrame:
        tokenized = self.tokenizer(
            list(batch["prompt"]), return_tensors="pt"
        )
        input_ids = tokenized.input_ids.to(self.model.device)
        attention_mask = tokenized.attention_mask.to(self.model.device)
```

```
        gen_tokens = self.model.generate(
            input_ids=input_ids,
            attention_mask=attention_mask,
            do_sample=True,
            temperature=0.9,
            max_length=100,
            pad_token_id=self.tokenizer.eos_token_id,
        )
        return pd.DataFrame(
            self.tokenizer.batch_decode(gen_tokens), columns=["responses"]
        )
```

All that is left is to run the `map_batches` method on the dataset in Listing 8-18. We specify that we want to use one GPU for each Ray Actor that will be running our callable class.

Also notice that we repartition the dataset into 100 partitions before mapping batches. This is to make sure there will be enough parallel tasks to take advantage of all the GPUs. 100 is an arbitrary number. You can pick any other numbers as long as it is more than the number of available GPUs in the cluster.

Listing 8-18. Getting batch predictions by applying `map_batches` to your Dataset

```
predictions = (
    ds
    .repartition(100)
    .map_batches(
        PredictCallable,
        batch_size=4,
        fn_constructor_kwargs=dict(
            model_id=model_id, revision=revision
        ),
        batch_format="pandas",
        compute=ray.data.ActorPoolStrategy(),
        num_gpus=1,
    )
)

predictions.take_all()
```

After map_batches is done, we can view our generated text using take_all().
Noticeably, we are not using a Predictor here. This is because Predictors are mainly
intended to be used with a Checkpoint, which we don't for this example.

Running Online Model Serving

The last aspect of Ray we want to demonstrate in this extended LLM example is how
to serve models like GPT-J-6B. To do this in Listing 8-19, we have to define a callable
class that will serve as a Ray Serve deployment (serve.deployment). At runtime, a
deployment consists of a number of replicas, which are individual copies of the class or
function that are started in separate Ray Actors. The number of replicas can be scaled up
or down to match the incoming request load.

We make sure to set the deployment to use one GPU by setting num_gpus accordingly.
We load the model in __init__, which will allow us to save time by initializing a model
just once and then use it to handle multiple requests.

Listing 8-19. Defining a Serve deployment using the serve.deployment
decorator on a Python class

```
import pandas as pd

from ray import serve
from starlette.requests import Request

@serve.deployment(ray_actor_options={"num_gpus": 1})
class PredictDeployment:
    def __init__(self, model_id: str, revision: str = None):
        from transformers import AutoModelForCausalLM, AutoTokenizer
        import torch

        self.model = AutoModelForCausalLM.from_pretrained(
            model_id,
            revision=revision,
            torch_dtype=torch.float16,
            low_cpu_mem_usage=True,
            device_map="auto",  # automatically makes use of all GPUs
            available to the Actor
        )
```

```python
        self.tokenizer = AutoTokenizer.from_pretrained(model_id)

    def generate(self, text: str) -> pd.DataFrame:
        input_ids = self.tokenizer(text, return_tensors="pt").input_ids.to(
            self.model.device
        )

        gen_tokens = self.model.generate(
            input_ids,
            do_sample=True,
            temperature=0.9,
            max_length=100,
        )
        return pd.DataFrame(
            self.tokenizer.batch_decode(gen_tokens), columns=["responses"]
        )

    async def __call__(self, http_request: Request) -> str:
        json_request: str = await http_request.json()
        prompts = []
        for prompt in json_request:
            text = prompt["text"]
            if isinstance(text, list):
                prompts.extend(text)
            else:
                prompts.append(text)
        return self.generate(prompts)
```

We can now bind the deployment with our arguments, and use ray.serve.run to start it. To run this script outside a Jupyter notebook, the recommended way is to use the serve run CLI command. In this case, you would remove the serve.run(deployment) line from Listing 8-20, and instead start the deployment by calling serve run <filename>:deployment from the command line.

Listing 8-20. Starting a Ray serve deployment by binding and running it

```
deployment = PredictDeployment.bind(
    model_id=model_id, revision=revision
)
serve.run(deployment)
```

Let's try submitting a request to our deployment in Listing 8-21. We will use the same prompt as in Listing 8-16 in the batch inference section, and send a POST request to localhost:8000. The deployment will generate a response and return it.

Listing 8-21. Getting predictions from our Serve deployment in Python

```
import requests

# reuse the same prompt as before
sample_input = {"text": prompt}

output = requests.post(
    "http://localhost:8000/", json=[sample_input]
).json()
print(output)
```

This wraps up our extended LLM example, discussing fine-tuning and inference with Ray. It's important to emphasize that throughout this example, we have been using a single Python script and a single distributed system in Ray to handle all the heavy work. The remarkable thing is that you can take this script and scale it out to a large cluster, utilizing CPUs for preprocessing and GPUs for training. You can easily configure the deployment separately by adjusting the parameters in the scaling configuration and similar options within the script. It's worth noting that this level of flexibility is not the standard. In practice, data scientists often find themselves using multiple frameworks for different tasks. For instance, they might use one framework for data loading and processing, another for training, and yet another for serving the trained models.

An Overview of Ray's Integrations

Next, we will explore the extensive range of integrations available for Ray. We will discuss this ecosystem from the perspective of Ray to better understand it within the context of a typical AI workflow. We will point you to additional resources when necessary, as we can't give any concrete code examples here.

Going in the order we introduced the Ray AI libraries earlier, let's quickly summarize the most common data formats supported by Ray Data in Table 8-2. Note that there is much more to be said about loading data with Ray Data, such as reading and writing to databases or cloud storage solutions. You can learn more about this topic from the Ray documentation.[16]

Table 8-2. *Supported data formats and third-party integrations of Ray Data*

Integration	Type	Description
Text, binary, image, CSV, JSON	Basic data formats	Ray Data has support for loading and storing many common data types
NumPy, Pandas, Arrow, Parquet	Advanced data formats	Ray Data supports with standard ML data structure libraries such as Pandas, but also supports many other formats such as Parquet
Spark, Dask, Mars, Modin	Advanced third-party integrations	Ray integrates with many data processing frameworks via community-sponsored integrations

Next, let's briefly list the most common integrations for the two Ray libraries dedicated to model training, namely, Ray Train and RLlib, in Table 8-3.

---

[16] Ray - Loading Data documentation, https://docs.ray.io/en/latest/data/loading-data.html

Table 8-3. *Available integrations for Ray Train and RLlib*

Integration	Type
TensorFlow, PyTorch, XGBoost, LightGBM, Horovod, Keras	Train integrations maintained by the Ray team
Scikit-learn, HuggingFace, Lightning	Train integrations maintained by the community
TensorFlow, PyTorch, OpenAI gym	RLlib integrations maintained by the Ray team
JAX, Unity	RLlib integrations maintained by the community

The integrations available for Ray Tune roughly split into two categories, namely, HPO libraries and experimentation tracking tools, as you can see in Table 8-4.

Table 8-4. *Integrations available in the Ray Tune ecosystem*

Integration	Type
Optuna, Hyperopt, Ax, BayesOpt, BOHB, Dragonfly, FLAML, HEBO, Nevergrad, SigOpt, skopt, ZOOpt	HPO library integration
TensorBoard, MLflow, Weights & Biases, CometML	Logging and experiment tracking integration

Lastly, let's quickly summarize the integrations available for Ray Serve, which split into integrations with model serving and observability frameworks (Table 8-5).

Table 8-5. *Integrations available for Ray Serve*

Integration	Type
FastAPI, Flask, Streamlit, Gradio	Serving frameworks and applications
Arize, Seldon Alibi, WhyLabs	Explainability and observability

Let's summarize all the integrations mentioned in this chapter in one concise diagram. In Figure 8-7 we list all integrations currently available throughout Ray.

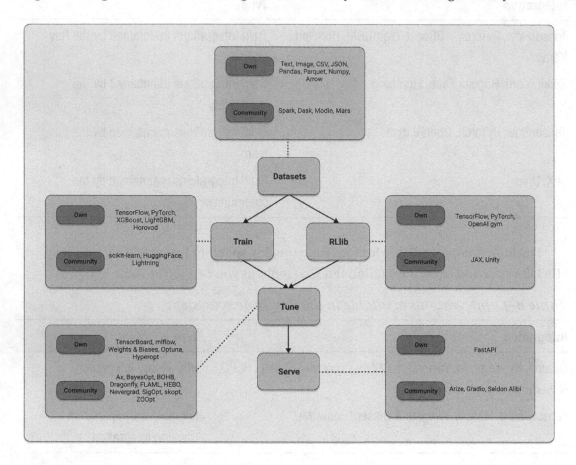

Figure 8-7. *Ray AI libraries and their ecosystem of integrations*

How Ray Compares to Related Systems

Now that you have gained a deeper understanding of Ray and its libraries, this section is an appropriate place to compare what Ray offers with similar systems. Ray's ecosystem is multifaceted, viewed from various perspectives, and utilized for different purposes. This means that many aspects of Ray can be compared to other tools available in the market. We will also discuss how to integrate Ray into complex workflows within existing machine learning platforms.

We haven't directly compared Ray with other systems yet. Given Ray's flexibility and its many components, it can be compared to different types of tools within the wider machine learning ecosystem. Let's begin by comparing Ray to more obvious candidates, specifically Python-based frameworks for cluster computing.

Distributed Python Frameworks

When it comes to distributed computing frameworks that provide comprehensive Python support and don't tie you to a specific cloud platform, the top contenders are Dask, Spark, and Ray. Although these frameworks have some technical and performance variations that depend on the specific context, it's most useful to compare them based on the types of workloads you intend to run. Table 8-6 outlines the comparison of the most common workload types.

Table 8-6. *Distributed Python workloads*

Workload Type	Dask	Spark	Ray
Deep Learning	Supported, not first-class	Supported, not first-class	First-class support
Structured data processing	First-class support	First-class support	Supported, not first-class
Low-level parallelism	First-class support using tasks	None	First-class support using tasks and actors

Ray AI Libraries and the Broader ML Ecosystem

Ray primarily focuses on AI compute, particularly by offering distributed training capabilities through Ray Train. However, it doesn't aim to cover every aspect of an AI workload. Instead, Ray integrates with tracking and monitoring tools for ML experiments and data storage solutions, rather than providing its own native solutions.

On the other hand, there are specific categories of tools where Ray can be seen as an alternative. For example, there are framework-specific toolkits like TorchX or TFX that tightly integrate with their respective frameworks. In contrast, Ray is framework-agnostic, avoiding vendor lock-in, and provides similar tooling.

It's also worth mentioning how Ray compares to specific cloud offerings. Some major cloud services offer comprehensive toolkits for handling ML workloads in Python. One notable example is AWS Sagemaker, which provides an all-in-one package that integrates well with the AWS ecosystem. Ray doesn't aim to replace tools like Sagemaker. Instead, it offers alternatives for compute-intensive components such as training, evaluation, and serving.

Ray can also serve as a viable alternative to ML workflow frameworks like KubeFlow or Flyte. Unlike many container-based solutions, Ray offers an easy-to-use Python API and native support for distributed data.

In some cases, the use of Ray may not be a straightforward decision, and it can be seen or used as both an alternative and a complementary component in the ML ecosystem. For example, as open source systems, Ray can be utilized within hosted ML platforms like SageMaker, or you can build your own ML platforms using them. Additionally, while Ray may not always directly compete with dedicated big data processing systems like Spark or Dask, Ray Datasets can often suffice for your processing requirements.

As we mentioned earlier, Ray's design philosophy revolves around expressing ML workloads in a single script and executing them on Ray as a unified distributed system. Since Ray handles task placement and execution on the cluster automatically, there's usually no need for explicit orchestration or complex integration with other distributed systems. However, it's important to note that this philosophy should be interpreted flexibly. Sometimes, multiple systems are necessary, or tasks need to be divided into different stages. In such cases, dedicated workflow orchestration tools like Argo or AirFlow can be valuable when used in combination with Ray. For example, you might choose to incorporate Ray as a step within the Lightning MLOps[17] framework.

How to Integrate Ray into Your ML Platform

To build your own ML platform and integrate Ray with other ecosystem components, the core of your system would consist of a set of Ray Clusters, each handling different tasks. For example, one cluster could handle preprocessing, training a PyTorch model, and running inference, while another cluster could focus on batch inference and model

---

[17] Lightning Ai, `https://lightning.ai/`

serving. To meet your scaling requirements, you can utilize the Ray Autoscaler and deploy the entire system on Kubernetes with KubeRay. Additionally, you can enhance this core system by incorporating other components based on your needs, such as

- Including other compute steps like Spark for data-intensive preprocessing tasks.

- Utilizing a workflow orchestrator such as AirFlow, Oozie, or SageMaker Pipelines to schedule and create your Ray Clusters, as well as run Ray apps and services. Each Ray app can be part of a larger orchestrated workflow, connecting with other components like a Spark ETL job.

- Creating Ray clusters for interactive use with Jupyter notebooks, which can be hosted on platforms like Google Colab or Databricks Notebooks.

- Integrating with feature stores like Feast or Tecton, as Ray Train, Datasets, and Serve offer integration with such tools.

- Leveraging experiment tracking and metric stores by integrating Ray Train and Tune with platforms like MLflow and Weights & Biases.

- Storing and retrieving data and models from external storage solutions like S3, as illustrated in the diagram.

By combining Ray with these additional components, you can build a comprehensive ML platform tailored to your specific requirements, as illustrated in Figure 8-8.

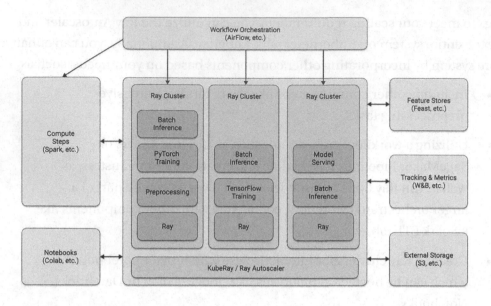

Figure 8-8. *Building an ML platform with Ray*

Summary

In this chapter, we've explored how the various Ray higher-level libraries were built for AI workflows. You've gained a deep understanding of the key concepts that enable you to build scalable ML projects, from experimentation to production. We've highlighted the use of Ray Datasets for stateless computations like feature preprocessing and demonstrated how Ray Train and Tune handle stateful computations like model training. Ray excels at seamlessly combining these types of computations in complex AI workloads and scaling them out to large clusters. Deploying your Ray ML projects is effortless using Ray Serve.

We've also introduced you to Ray's ecosystem, empowering you to run your own experiments alongside the tools you're currently using or planning to use in the future. Additionally, we've discussed the limitations of Ray and compared it to various related systems. We've explored how you can leverage Ray in conjunction with other tools to enhance or build your own ML platforms.

With this knowledge, you are well-equipped to embark on your own Ray experiments and utilize the tools that best suit your needs. You have the flexibility to integrate Ray with other tools and expand your ML capabilities as desired.

## CHAPTER 9

The Future of MLOps

As AI/ML matures and businesses increasingly rely on them for business competitive advantages, such as enhancing customer experience and driving growth, a robust MLOps infrastructure becomes critical. Just as data infrastructure became an essential for managing and analyzing data for data-driven companies, MLOps has transformed into the essential backbone for effectively developing, deploying, managing, and monitoring AI/ML models in production at scale.

MLOps Landscape

While AI/ML has immense potential to transform businesses, companies need to carefully consider the challenges and invest in the necessary infrastructure to unlock its full potential. As they are turning their AI/ML strategies into execution, companies will increasingly rely on MLOps or ML infrastructure to operationalize AI/ML models and scale their AI/ML initiatives effectively and efficiently.

The MLOps landscape is now brighter and more clear than ever before, having moved firmly beyond the hype cycle and benefiting from the emergence of large language models. This convergence fuels optimism for a future where MLOps play an even more critical role in companies.

ML Development Lifecycle

It is widely recognized that ML development differs from traditional software development, primarily due to the nature of machine learning and the additional critical component known as data. Data serves as an important input element in ML development, injects significant complexity into the process in terms of model versioning and reproducibility. The ML development consists of a series of stages, as

© Hien Luu, Max Pumperla and Zhe Zhang 2024
H. Luu et al., *MLOps with Ray*, https://doi.org/10.1007/979-8-8688-0376-5_9

depicted in Figure 9-1. While it follows a distinct pattern, it is inherently an iterative process and requires a substantial amount of experimentation to produce high performance models.

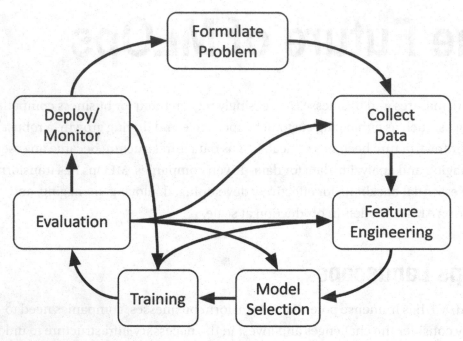

Figure 9-1. *ML development, an iterative process*

ML development is often described as a team sport that consists of multiple personas, including data engineer, data scientist, ML engineer, business stakeholders, MLOps engineer, and more. As such, effective collaboration, communication, and coordination among these roles are essential for the success of ML development.

In recent years, the ML community has made significant strides in gaining a deeper understanding of the complexities and challenges involved in ML development. This has been driven by a combination of factors, including advances in research, real-world experience, and the sharing of knowledge and resources. There are now abundant resources available to support ML developers at all levels, from beginners to experts. These include blogs, books, conferences, and courses, all of which provide valuable insights and knowledge to help developers improve their skills and stay up-to-date with the latest development in the field.

ML Infrastructure Architecture

The ML infrastructure is a dynamic and ever-evolving tapestry of components. At the same time, there is no single tool that can address all the needs of the various phases of ML development in a perfect way. For organizations that are scaling up their ML investment and initiatives, it has become a necessity for them to invest in their ML infrastructure. The good news is that those who have made such an investment found the benefits outweighed the costs in a short span of time.

As an ML community, after a few years of gaining experience with applying ML in the enterprise settings, best practices and the canonical stack of ML systems have emerged. As an example, a set of ML infrastructure capabilities[1] is captured in this blueprint by the AI Infrastructure Alliance, as depicted in Figure 9-2. The aim here is to help companies to speed up the process of building or putting together their ML infrastructure and to avoid common pitfalls.

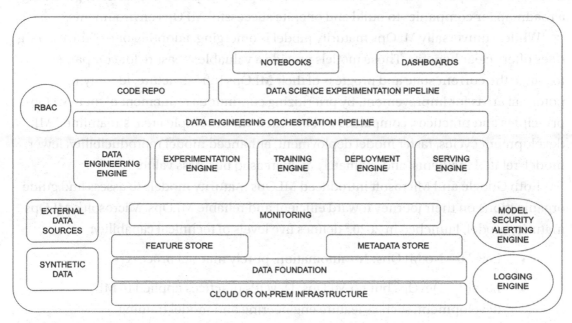

Figure 9-2. *ML development, an iterative process*

[1] AI Infrastructure Ecosystem, 2022, https://ai-infrastructure.org/ai-infrastructure-ecosystem-report-of-2022/

The above blueprint serves as a reference guide, outlining the essential capabilities for building a robust ML infrastructure. While the ideal layout serves as a valuable starting point, each company's unique needs and operating characteristics will influence its specific implementation. Factors like team size and their experience, development methodologies and process, and preferred tech stack will shape the chosen tools, deployment techniques, monitoring strategy and rigor, and more. Companies should feel empowered to tailor the blueprint to their specific context.

MLOps Maturity Model

Mirroring the software maturity model, MLOps is now establishing its own framework for assessing an organization's practices and progress toward efficient, reliable, and high-quality ML operations. This emerging model, largely driven by cloud vendors with extensive ML experience, aims to illuminate MLOps principles and practices, providing a roadmap for companies to build and operate successful MLOps environments.

While a universally MLOps maturity model is emerging, adopting one of the existing ones offers clear benefits. Those models provide a valuable measure for companies to gauge the current state and progress of their MLOps infrastructure, identifying potential areas for improvement. By prioritizing continuous refinement of their MLOps principles and practices, companies will benefit from a multiple area: streamlined ML development cycles, faster model deployment, enhanced model reproducibility, fewer model-related incidents, and ultimately an increased business value.

Both Google and Microsoft introduced MLOps maturity models to assess and guide organizations on their journey toward efficient and reliable MLOps. Microsoft's MLOps maturity model, launched in 2020,[2] defines five levels of technical capability:

- Level 0 – No MLOps: No automation, purely manual process

- Level 1 – DevOps but no MLOps: DevOPs practices applied to ML development, such as feature engineering and model training pipelines

- Level 2 – Automated training: Automation of the ML training pipeline

---

[2] Microsoft, Machine Learning operations maturity model, 2020, https://learn.microsoft.com/en-us/azure/architecture/ai-ml/guide/mlops-maturity-model

- Level 3 – Automated model development: Automation of the model deployment to production

- Level 4 – Full MLOps automated operations: end-to-end automation of the entire lifecycle, including continuous model retraining and deployment

Google's MLOps maturity model, also released in 2020,[3] takes a more streamlined approach with three maturity levels:

- Level 0 – Manual process: Manual development and deployment of ML models

- Level 1 – ML pipeline automation: Continuous model training through automated pipelines

- Level 2 – CI/CD pipeline automation: Rapid and reliable model updates in production via a robust CI/CD system

Despite the difference in the number of levels, both maturity models share a common focus on automation as the key driver of MLOps maturity. Automating ML development steps like feature generation, model development, training, testing, deployment, monitoring leads to reliable, repeatable, scalable, and reproducible pipelines that significantly increase model development velocity. This, in turn, empowers organizations to unlock the great potential of their ML initiatives.

MLOps Solution Landscape

Over the last five years, there has been a proliferation of MLOps solutions, emerging from both the open source community and the vendor space. This landscape has become expansive and is continuously evolving. The landscape captured on AI Infrastructure Landscape contains over 60 companies.[4] While this abundance of options provides companies with a variety of choices to choose from, it can also lead to confusion and time consuming when trying to nail down specific options or determine where to initiate the decision-making process.

---

[3] Google, MLOps: Continuous delivery and automation pipelines in machine, learning, 2020, https://cloud.google.com/architecture/mlops-continuous-delivery-and-automation-pipelines-in-machine-learning

[4] AI Infrastructure Landscape, https://ai-infrastructure.org/ai-infrastructure-landscape/

Companies tend to shy away from vendor lock-ins whenever possible, but at the same time, they might not have sufficient talent and resources to take the DIY path. The majority of the companies are opting for best-of-breed solutions from in-house, open source, and vendor solutions.

AI/ML Landscape

The AI/ML landscape is shifting. Once primarily focused on research and development, the focus is now turning toward operationalizing these models at scale. As AI/ML become increasingly mainstream, their impact on business is undeniable. It is not surprising that AI/ML adoption has steadily increased over the past few years, and will continue to grow in the foreseeable future. More and more companies are moving beyond pilot projects to deploying AI/ML at scale across multiple departments. This indicates companies have more confidence in its value and ability to deliver ROI. According to the "2023 State of Data + AI" inaugural report from Databricks, a survey across their large customer base, they found there was an increase of 411% of YoY growth in deploying more models in production.[5]

Advancements in deep learning and natural language processing are opening up new possibilities for applications across various industries, from transportation, healthcare, to finance and entertainment, and more. In transportation, self-driving cars powered by deep learning algorithms are navigating city streets with increasing precision, promising safer and more efficient commutes. In the financial sector, Natural Language Processing (NLP) is assisting in financial analysis, extracting insights from market data and generating informed investment recommendations.

For enterprises, there are plenty to be excited about when it comes to AI/ML's potential to transform their businesses. However, there are several key concerns when it comes to integrating this powerful technology at their companies:

- Explainability and interpretability: Deep learning models tend to be black boxes, and it is not easy to understand how they arrive at their predictions. Not being able to explain the internal logic of complex models raises concerns about bias and fairness when dealing with sensitive data like consumer-related information.

---

[5] 2023 State of Data + AI, 2023, www.databricks.com/resources/ebook/state-of-data-ai

- Data security and privacy: Protecting misuse of customer and sensitive data is extremely critical for building trust with customers. Companies need robust data and AI security measures and ensure their AI/ML applications comply with various regulations and privacy laws.

- Talent shortage: Implementing AI/ML projects and managing them in production requires a distinct skill set that is presently in high demand. The talent shortage poses challenges for businesses to find qualified personnel, and this can slow down adoption.

- Cost and ROI: Implementing AI/ML projects often requires a sizable financial investment in infrastructure, talent acquisition, and ongoing maintenance. The ROI of these projects may not be immediately apparent, and organizations need to be mindful of both short-term gains and long-term strategic advantages.

Generative AI

The next frontier in AI/ML is Generative AI, which refers to AI systems designed to create or generate new data or content, rather than analyzing existing data to find patterns, as in traditional AI. The types of content it can generate include sophisticated text, images, videos, audio, code, and more, at the level that mimics human ability. This opens up a whole new world of possibilities, from designing innovative products and materials to crafting personalized experiences and even generating art and code, or composing music.

Generative AI has the potential to change the anatomy of work and has a huge impact on productivity across industries and business functions. McKinsey Global Institute estimates that Generative AI could add the equivalent of $2.6 trillion to $4.4 trillion in global corporate value annually across 63 use cases analyzed.[6] The large and diverse set of use cases in which they estimate that Generative AI will raise productivity, including providing support interactions with customers, generating creative content for marketing sales, and drafting software code based on natural-language prompts, and more.

---

[6] McKinsey & Company, The economic potential of generative AI, 2023, www.mckinsey.com/capabilities/mckinsey-digital/our-insights/the-economic-potential-of-generative-ai-the-next-productivity-frontier#

Foundation Models

The powerful and diverse generative capabilities mentioned above are powered by foundation models, which is a term coined by Stanford researchers to introduce a new category of ML models.[7] These models are based on deep neural networks and typically trained in a self-supervised way on massive amounts of unstructured datasets, often sourced from the Internet, including books, articles, websites, and other publicly available media. The Stanford researchers see foundation models as a paradigm shift; instead of one model built solely for one task, foundation models can be adapted across a wide variety of different tasks, as depicted in Figure 9-3. The adaptability turns foundation models in general into highly potent and versatile tools, similar to a Swiss Army knife.

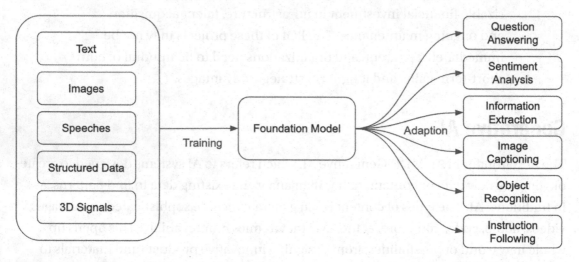

Figure 9-3. *A foundation model trained with data from various modalities and adapted to a wide range of tasks[8]*

Large Language Models (LLMs)

Large language models (LLMs) are a type of foundation models trained on a large volume of text data to primarily focus on natural language understanding and text generation. The "large" in their name not only denotes the vast training datasets but also their neural network size in terms of billions of parameters and large amount of

---

[7] Rishi Bommasani, Percy Liang, Reflections on Foundation Models, https://hai.stanford.edu/news/reflections-foundation-models

[8] Center for Research on Foundation Models, On the Opportunities and Risks of Foundation Models, 2022, https://arxiv.org/pdf/2108.07258.pdf

computational power required to train these models. These state-of-the-art language models are trained using a specific neural network architecture called Transformer and are able to demonstrate exceptional language capabilities, including text generation, translation, summarization, answering questions, and more. The transformer architecture, introduced by the seminal paper from Google called "Attention is all you need," has had a transformative impact on the natural language processing field. A key concept of the transformer architecture is self-attention. This is what allows LLMs to understand relationships between words and gain sophisticated capabilities to parse language in an efficient manner.

Although LLMs have been seen to exhibit capabilities similar to humans, such as generating highly fluent and coherent text with specific styles and tones, translate text from one language to another, summarize a large body of text in a concise and clear manner, at the most basic level, LLMs are generative mathematical models of statistical distribution of tokens in the vast public corpus of human-generated text.[9] Their ability to sample from this distribution makes them powerful next-token generators, enabling them to produce coherent and contextually relevant text based on learned language patterns.

Generative Pre-trained Transformer (GPT) models are a specific category of LLMs. The term GPT is often referred to a series of LLMs specifically developed by OpenAI, and the underlying architecture of these models is the transformer architecture. As of this writing, the latest version of GPT model GPT-4 was launched in March 2023. GPT-4 is a large multimodal model capable of processing image and text inputs and producing text outputs. It was trained using both publicly available data from the Internet and data licensed from third-party providers. In the blog post "GPT-4," they touted GPT-4 while less capable than humans in many real-world scenarios, exhibits human-level performance on various professional and academic benchmarks.[10]

[9] Murray Shanahan, Talking About Large Language Models, 2022, https://arxiv.org/pdf/2212.03551.pdf

[10] OpenAI, GPT-4, 2023, https://openai.com/research/gpt-4

AI Assistants

ChatGPT is a specialized application of GPT that has been fine-tuned for conversational-based tasks, making it a valuable tool for chatbots, virtual assistants, and other interactive applications. It is designed to provide responses that align with the ongoing dialogue by maintaining context in a conversation, allowing for more coherent and contextually relevant interactions.

ChatGPT has long enjoyed the first-mover advantage since it was launched in 2022. It is no surprise that ChatGPT alternatives have emerged. The notable ones are Bard from Google, Bing AI Chat from Microsoft, and Claude 2 from Anthropic. It is very likely there will be more alternatives available in the near future.

GitHub Copilot and Amazon CodeWhisperer are coding assistants that leverage LLM capabilities specially for software development. They are designed to assist developers and boost their productivity by providing real-time code suggestions and recommendations within an integrated development environment.

For image generation, some of the popular assistants are Stable Diffusion, DALL-E 3, and Midjourney.

A good way to visualize the relationship between generative AI, foundation models, large language models, and AI assists is through the diagram in Figure 9-4.

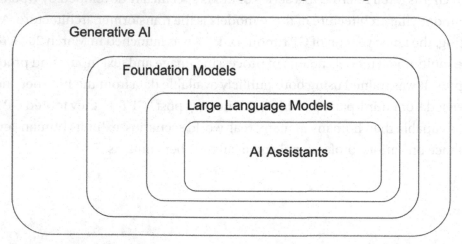

Figure 9-4. *AI topology and how the various pieces fit together*

Responsible AI

According to Sam Altman, OpenAI CEO, AI will be the single most powerful and transformative technology of this era, an assessment he made at an interview with *Time* magazine.[11] At the same time, this powerful and transformative technology is far from perfect and has a few risks and limitations. Among them are hallucination, misinformation, bias, and security risks:

- Hallucination: Generative AI models can generate incorrect information that is not based on any factual data. In other words, they fabricate facts or present inaccuracies as truths. This is an ongoing area of research and development. Robust evaluation and continuous monitoring are two of the many best practices that organizations should invest in while building generative AI applications.

- Misinformation: Generative AI models can be used to easily and quickly generate persuasive and intentional misinformation to deceive or mislead people. Examples of misinformation include fake celebrity advertisements, fabricated political speeches, and more. Misinformation has a profound thread to countries around the world. A collaborative approach has been advocated to integrate the generative AI advancements with regulatory strategies and media literacy education to combat this broad societal issue.[12]

- Bias: Generative AI models are trained on large datasets that may contain societal bias, and therefore they have the potential to reinforce and reproduce biases. Examples of approaches to address bias include careful data curation to ensure diverse and representative training data, continuous bias detection and evaluation, establish clear ethical guidelines on the use of these models, and more.

[11] Simmone Shah, Sam Altman on OpenAI, Future Risks and Rewards, and Artificial General Intelligence, 2023, `https://time.com/6344160/a-year-in-time-ceo-interview-sam-altman/`

[12] Mohamed Shoaib, Zefan Wang, Milad Ahvanooey, Jun Zhao, Deepfakes, Misinformation, and Disinformation in the Era of Frontier AI, Generative AI, and Large AI Models, 2023, `https://arxiv.org/pdf/2311.17394.pdf`

- Security risks: A prompt injection risk is where malicious prompts
  are injected to manipulate the model into generating unintended
  or harmful responses. Another security risk is jailbreaking, which
  is a technique to bypass the internal safeguards of generative AI
  models. These risks highlight the need for robust security measures
  to safeguard against potential exploits.

It is encouraging to see leading large AI companies and governments around the world collaborating on a plan for safety testing of the generative AI models. This will help establish a shared understanding of their capabilities and potential risks posed by these models. At the 2023 AI Safety Summits in London, an AI Safety Institute[13] was launched and received warm welcomes and strong support from numerous world leaders and heads of major AI companies.

Artificial General Intelligence (AGI)

There has been an uptick in discussion about or mentioning of the term AGI since the release of GPT-4 in March 2023. The phrase "Artificial General Intelligence" was re-introduced by Shane Legg about 20 years ago to vaguely describe the generality that AI systems don't yet have[14] while having a conversation with his fellow researcher about their book of essays about AI.

AGI is one of the most controversial topics in the AI research community and a big part of this problem is due to the fact that most AI experts have their own definition about what AGI is.

In November of 2023, a team of Google Deepmind researchers proposed a framework for classifying the capabilities and behavior of AGI concept in a paper.[15] They hope this framework will provide a common language to compare models, assess risks,

---

[13] UK Prime Minister's Office, Prime Minister launches new AI Safety Institute, 2023, www.gov.uk/government/news/prime-minister-launches-new-ai-safety-institute

[14] William Heaven, Artificial general intelligence: Are we close, and does it even make sense to try?, 2020, www.technologyreview.com/2020/10/15/1010461/artificial-general-intelligence-robots-ai-agi-deepmind-google-openai/

[15] Meredith Morris, Jascha Sohl-dickstein, Noah Fiedel, Tris Warkentin, Allan Dafoe, Aleksandra Faust, Clement Farabet, Shane Legg, Levels of AGI: Operationalizing Progress on the Path to AGI, 2023, https://arxiv.org/pdf/2311.02462.pdf

and measure progress along the path to AGI. It includes a set of principles for a clear and operationalizable definition of AGI and introduces six levels of AGI ontology, similar to the levels of autonomous driving.

The Rise of LLMOps

The advent of LLMs introduces a new class of applications known as GenAI applications. As organizations explore the potential of this vast AI frontier, they face the challenges of operationalizing this new class of applications, primarily centered around LLMs.

LLMs are powerful and versatile, at the same time, they pose challenges related to using, managing, and deploying them at scale. These challenges are mainly stemming from their complexity, size, coupled and text-based inputs and outputs. Consequently, a new discipline has emerged to address these complexities, recognized as LLMOps (Language Model Operations).

Fundamentally similar to MLOps, LLMOps is the discipline that focuses on the tools, principles, and best practices for operationalizing the lifecycles of GenAI applications. It specifically focuses on LLM applications interacting with either closed source or open source LLMs. Moreover, its overarching goal aligns with that of MLOps: to accelerate the development and deployment velocity of LLM applications.

LLM Applications Archetypes

To understand the composition and challenges inherent with LLMOps, one must initially understand the diverse archetypes of LLM applications. These archetypes are not mutually exclusive; instead, they are additive. This implies each archetype can be integrated into the progressively more sophisticated ones.

Prompt Engineering Application Archetype

This LLM application archetype is the simplest among the three. At the high level, this application archetype simply sends instructions, known as prompts, to LLMs to get them to generate the needed content, as depicted in Figure 9-5. The art of crafting the most suitable prompt using various strategies and tactics for getting optimal results from LLMs is known as promoting engineering. Examples of popular vendor provided LLMs include GPT-4 from OpenAI, Gemini from Google, Claude 2 from Anthropic, Coral from Cohere, and more.

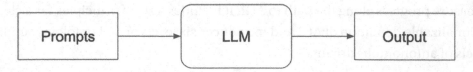

Figure 9-5. *Prompt engineering application archetype*

This application archetype leverages well-crafted prompts to perform a diverse range of natural language processing (NLP) tasks. Examples include text summarization, information extraction, text classification, code generation, and more.

From the LLMOps standpoint, the primary concerns regarding operationalizing this type of application center around prompt development and management. This includes prompt development and test, versioning, logging, monitoring, and thorough testing.

Retrieval Augmented Generation (RAG) Application Archetype

RAG (Retrieval Augmented Generation) has become one of the most popular methods for building LLM applications. It combines the generative power of LLMs with the factual accuracy of external knowledge sources to create more informed and reliable output.

LLMs often suffer from factual inconsistencies and lack of context. They primarily rely on their internal statistical models, which can lead to creative but potentially inaccurate outputs. The RAG method helps with addressing this gap by injecting factual grounding and external data. Essentially, it is a way to give LLMs a "boost" with real-world knowledge.

The RAG method's rise to popularity because it has been researched[16] and found to be an effective method to address the aforementioned gap. This is critical for LLM applications that need domain-specific information that are behind corporate firewalls, and that might change frequently. This is specially relevant for enterprise RAG-based LLM applications that need incorporating their company's proprietary data.

---

[16] Aleksandra Piktus, Fabio Petroni, Vladimir Karpukhin, Naman Goyal, Heinrich Kuttler, Mike Lewis, Wen-tau Yih, Tim Rocktaschel, Sebastian Ridel, Douwe Kiela, Retrieval-Augmented Generation for Knowledge-Intensive NLP tasks, 2021, https://arxiv.org/pdf/2005.11401.pdf

The widely applicable RAG method is particularly well-suited for LLM applications that require high levels of accuracy and domain-specific information, including question and answering systems, document summarization systems, content creation, chatbots, legal research systems, and more.

At the high level, the architecture of RAG-based applications includes components for transforming the domain data into vector embedding, loading them and the associated metadata into a vector database, querying the vector database and combing the found context with the prompt before sending them over to the LLMs. Figure 9-6 only depicts the flow during the content generation phase.

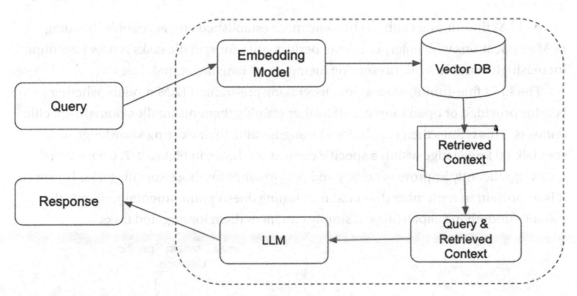

Figure 9-6. *RAG-based content generation flow*

From the LLMOps standpoint, the primary concerns regarding operationalizing RAG-base applications center around the following major parts:

- Automated pipelines chunking the data, generating embeddings and uploading them to a vector DB in a reliable and efficient manner

- Productionalizing and monitoring a scalable, reliable, high performant vector DB

- Logging and monitoring of the retrieval results from the vector DB

Fine-Tuning Application Archetype

While LLMs trained on massive datasets of text and have a broad range of language understanding and language capabilities, they often fall short in achieving acceptable levels of performance when applied to domain-specific tasks. LLM fine-tuning methods have become an important tool to adapt LLMs into specialized models and unlock their true potential to excel in targeted and domain-specific tasks. The popularity of this method has dramatically increased with the increased availability of smaller, but effective LLMs in the second half of 2023, such as Llama 2, Mistral, Dolly, and Falcon, among others.

As LLM fine-tuning methods become more established and accessible, building LLM applications with enterprise-level performance for specific tasks is now becoming increasingly feasible while preserving their general language capabilities.

The LLM fine-tuning process involves taking pre-trained LLM models, whether vendor provided or open source, and further training them on smaller, domain-specific datasets. This refines their capabilities by augmenting their existing knowledge with specialized knowledge within a specific context, as shown in Figure 9-7. Consequently, it can significantly improve accuracy and performance for those specific tasks. However, it is important to remember that LLM fine-tuning doesn't fundamentally change the model's underlying capabilities; it simply enhances them for targeted tasks.

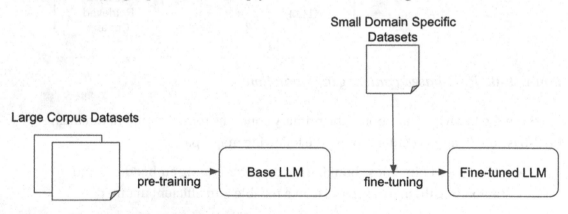

Figure 9-7.   *High-level flow of fine-tuning LLMs*

LLM fine-tuning, while powerful, can be computationally expensive, requiring multiple GPUs and large amounts of memory. Fortunately, the ML community has developed various optimization techniques to expedite training and ease the memory constraints. Among these, the Low-Rank Adaption (LoRA)[17] technique from the Parameter Efficient Fine-tuning (PEFT) family stands out for its effectiveness in reducing the number of parameters requiring updates during fine-tuning.

From the LLMOps standpoint, the primary concerns regarding operationalizing fine-tuning applications center around the following major parts:

- A standard, configurable, and repeatable workflow for fine-tuning LLMs

- Workflow users only need to provide the necessary input and specify a few configuration values, such as

  - The small and domain-specific dataset to fine-tune the base LLM

  - One or more base LLM models to fine-tuning

  - The validation dataset to evaluate the fine-tuned LLMs

  - Number of GPUs and memory size

  - Fine-tuning hyperparameters, such as learning rate, number of epochs

  - Fine-tuning technique and their respective configuration values

- Provisioning the compute cluster with sufficient compute and memory resources, and appropriate libraries for distributed parallel training

- The logging of training time, compute and memory usage efficiency, and the specified inputs and configuration values

- An LLM serving stack to host and serve open source or vendor provided fine-tuned LLMs

---

[17] Edward Hu, Yelong Shen, Philip Wallis, Zeyuan Zhu, Yuanzhi Li, Shean Wang, Lu Wang, Weizhu Chen, LoRA: Low-Rank Adaptation of Large Language Models, 2021, https://arxiv.org/abs/2106.09685

Model Training vs. LLM Fine-Tuning

While there are certainly similarities between the model training process and the LLM fine-tuning process, there are a few key differences worth noting:

- Memory requirements: Due to the large size of LLMs, fine-tuning them typically requires substantially more hardware memory, which can lead to higher costs for memory and GPU usage.

- Computational cost: As fine-tuning LLMs can be computationally intensive, it can require the use of powerful GPUs, which can be expensive to acquire and maintain.

- GPU shortage: The increasing demand for GPUs for LLM fine-tuning and serving has led to a shortage of GPUs, further exacerbating the cost issue and an increased in wait time in securing them.

A Combined Application Archetype

With maturing LLM technology and best practices, and as organizations gain deeper understanding and experience in building LLM applications, combining two or more of the above archetypes will become standard practice for designing sophisticated LLM applications that maximize LLM potential and add business value. This requires a comprehensive LLMOps stack, which we will explore next.

LLMOps Stack

While the LLMOps landscape is still nascent and undergoing continuous refinement, the application archetypes outlined offer valuable insights into the composition and essential components of a robust and comprehensive LLMOps stack.

In their 2023 blog post "Emerging Architectures for LLM Applications,"[18] the investment firm Andreessen Horowitz presented a fairly comprehensive reference architecture for the emerging LLM app stack, as shown in Figure 9-8.

---

[18] Matt Bornstein, Rajko Radovanovic, Emerging Architectures for LLM Applications, 2023, https://a16z.com/emerging-architectures-for-llm-applications/

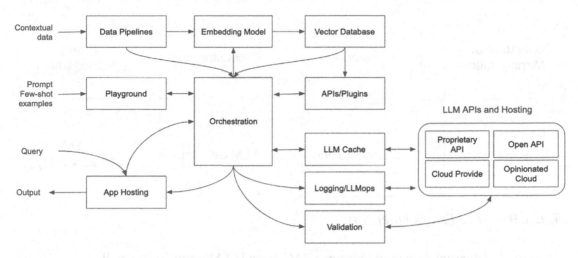

Figure 9-8. *Emerging LLM App Stack[17] from A16Z*

Though the blog primarily focuses on the LLM app stack, it implicitly highlights essential elements for an LLMOps stack supporting the canonical LLM application workflow. This encompasses data preprocessing and embedding generation, prompt construction and external knowledge retrieval, and prompt execution and LLM inference.

Industry leaders like LinkedIn and Uber, pioneers in the MLOps domain, are paving the way for generative AI platforms. They are extending their existing MLOps platforms to handle the unique needs of LLM applications. At the @Scale conference in July 2023, Min Cai, presenting about Uber's "Michelangelo ML Platform at Uber: Past, Present and Future"[19] revealed a few LLM related capabilities that will be added to their platform, such as API gateway, prompt engineering toolkit, hallucination detector, and more.

Given the rapidly evolving landscape of LLMs, most proposed LLM stacks will need to be revised shortly after they are created. The blueprint outlined in Figure 9-9 lays out essential components for an enterprise-ready LLM stack. Below, we delve into each component from an operationalization perspective, providing brief descriptions. By examining each component in detail, we can gain a deeper understanding of the complexities of operationalizing successful LLM applications in an enterprise setting.

---

[19] Min Cai, Michelangelo ML Platform at Uber: Past, Present and Future, 2023, www.youtube.com/watch?v=Z3-HddkYgyI

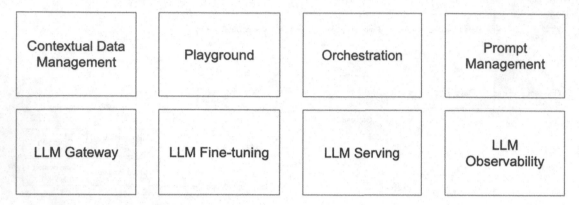

Figure 9-9. *LLM stack blueprint*

- Contextual data management: RAG-based LLM applications will rely heavily on the support from this component. Automating and orchestrating embedding generation and ingestion pipeline will be crucial. Additionally, the careful choice of a production-ready and scalable vector database for embedding storage and semantic search will be pivotal to the overarching architecture of RAG-based LLM applications.

- Playground: It is crucial for organizations to foster the spirit of LLM experimentation and exploration. Playground, an interactive environment for testing and experimenting with various LLMs. It serves as a valuable tool for education, and advancing the development and understanding of LLMs. The value LLMs may add to an organization's business depends on their knowledge and intuition around this technology.

- Orchestration: Sophisticated LLM applications will typically combine multiple capabilities, including knowledge retrieval and augmentation from various sources, prompt chaining, prompt execution, state management, and more. An orchestrator serves as the conductor, providing a cohesive framework and abstraction that simplifies LLM application development.

- Prompt management: Prompts are the primary means for interacting with LLMs, and they play a critical role in determining the quality of the models' responses. Effective prompt management is essential to ensure that they are carefully crafted, well-tested, and appropriately versioned and logged. This requires a comprehensive process and production-ready technology to manage prompts effectively.

- LLM gateway: To avoid vendor lock-in and to tap into the unique capabilities of multiple LLMs, LLM applications will often need to interact with multiple models. An LLM gateway plays a crucial role in facilitating this by acting as a central point of access. Beyond mere access, it provides important functionalities like logging, auditing, load balancing, rate limiting, and throttling, ensuring interaction with LLMs are managed effectively and efficiently. The LLM gateway enables organizations to build flexible and scalable LLM applications that can leverage the best of breed LLMs.

- LLM fine-tuning: Customization is often key to the success of enterprise LLM applications. Adapting LLMs to specific tasks through fine-tuning will become a common practice. However, this computationally intensive process can be expensive and time-consuming, requiring experimentation. To support effective and efficient fine-tuning of LLMs at scale, organizations need to have access to robust infrastructure and tools that simplify the process and reduce the associated costs. This includes hardware and software systems that can handle the computational demands of fine-tuning, as well as expert support to guide the process and optimize the outcomes.

- LLM serving: LLM serving commonly goes hands in hands with LLM fine-tuning, as it involves hosting, deploying, and serving the fine-tuned models. The LLM serving infrastructure plays a critical role in enabling the effective deployment and management of these models. It incorporates various optimization techniques during inference, including accelerating response time, boosting throughput, and maximizing compute efficiency. This ensures the fine-tuned LLMs can be effectively utilized in real-world LLM applications and delivers the desired performance and capabilities.

- LLM observability: This encompasses two key aspects: LLM evaluation and LLM monitoring. Evaluating LLMs can be challenging due to the complexity of assessing the quality of the text-based output. It requires a robust infrastructure and rigorous evaluation techniques to ensure models meet high standards of accuracy, fairness, and robustness in their responses. LLM monitoring is a continuous process of monitoring the performance of the models over time to ensure they are meeting the expected performance criteria. This includes monitoring of prompts provided to the models, their responses, and various operational metrics.

The specific LLM use cases and types of LLM applications being developed within an organization will guide the decision-making process in determining the degree of sophistications required for each of the components in the LLM stack. For instance, a company that is focused on developing a single, highly customized LLM application for a specific use case may require a more sophisticated LLM fine-tuning component, while a company that is working on a portfolio of broader-scope LLM applications may prioritize a more robust LLM serving infrastructure. Understanding the specific needs and priorities of the organization is crucial for making informed decisions about how to allocate resources and build a scalable, flexible, and effective LLM stack.

LLMOps is still a relatively nascent discipline. It is expected to continue to develop and evolve as LLMs and LLM applications become more prevalent in the industry.

Summary

As AI/ML matures and continues to deliver business values to organizations around the world, the future of MLOps is bright and exciting. MLOps has become a widely accepted discipline to help speed up the ML development velocity and bring models to production quicker, more reliably, and in a repeatable manner. With the emergence of MLOps tools, frameworks, and solutions from the open source and vendors, companies now have more choices than ever before in their effort to put together or build their ML infrastructure.

As organizations progress into the later phases of their MLOps adoption, it is important to adopt a MLOps maturity level framework to help assess and continuously improve their ML operations. This leads to a more efficient and effective deployment of the ML models, ultimately accelerating time to ROI from their ML initiatives.

The AI/ML landscape is expanding rapidly with the arrival of generative AI. This dynamic field involves using cutting-edge deep learning techniques to train LLMs with a massive amount of content from books, Internet, and other sources. These models can generate new content, such as text, images, code, and music, disrupting content creation in various fields. This exciting technology has the potential to dramatically accelerate innovation and increase productivity across various industries. However, along with the excitement, this technology also brings ethical considerations and challenges concerning potential for misuse, manipulation, and misinformation.

Organizations are exploring ways to harness the power of LLMs and actively learning about building and operationalizing LLM applications. Several common LLM application archetypes have emerged. Managing and employing LLMs requires new infrastructure and expertise. LLMOps, a dedicated set of practices and tools tailored for LLMs, is a burgeoning field crucial for building and operationalizing LLM applications. The ML community and MLOps vendors are actively developing blueprints and technology stack to help organizations tackle the challenges of navigating the complexity of those tasks.

Index

A

Adoption strategies
  AI/ML infrastructure, 44
  alignment, 45
  business domains, 46
  churn prediction, 47
  culture elements
    collaboration, 52, 53
    decision-making process, 52
    dimensions, 51
    execution velocity, 51
    risk tolerance level, 51
  DevOps/DataOps, 49, 50
  fraud detection, 47
  infrastructure, 45
  landscape, 63
  loan approval and credit scoring, 48
  maturity levels, 53–55
  Meta FBLearner, 70–74
  MLOps infrastructure, 45, 46
  model governance, 48
  online products, 66
  organizations, 74
  platforms/specialist tools, 64–66
  team members, 50
  technologies or infrastructure, 56–63
  Uber Michelangelo, 67–70
  vast and complex infrastructure, 67
Application programming
    interfaces (APIs)

computation, 253
functions, 251
GPU chips, 254, 255
Python code, 251, 252
Ray asynchronous computing, 252
task graph execution, 253
Artificial General Intelligence
    (AGI), 316
Artificial intelligence (AI)
  adoption strategies, 44
  AGI concepts, 316
  assistants, 314, 315
  foundation model, 312
  generative AI, 311
  key concerns, 310
  landscape, 310
  large language models (LLMs),
    312, 313
  risks and limitations, 315, 316
Asynchronous predictions, 35

B

Batch predictions vs. online
    prediction, 35

C

Canary model deployment, 173
Chat Generative Pre-trained Transformer
    (ChatGPT), 57

U, V, W, X

Printed in the United States
by Baker & Taylor Publisher Services

Printed in the United States
by Baker & Taylor Publisher Services